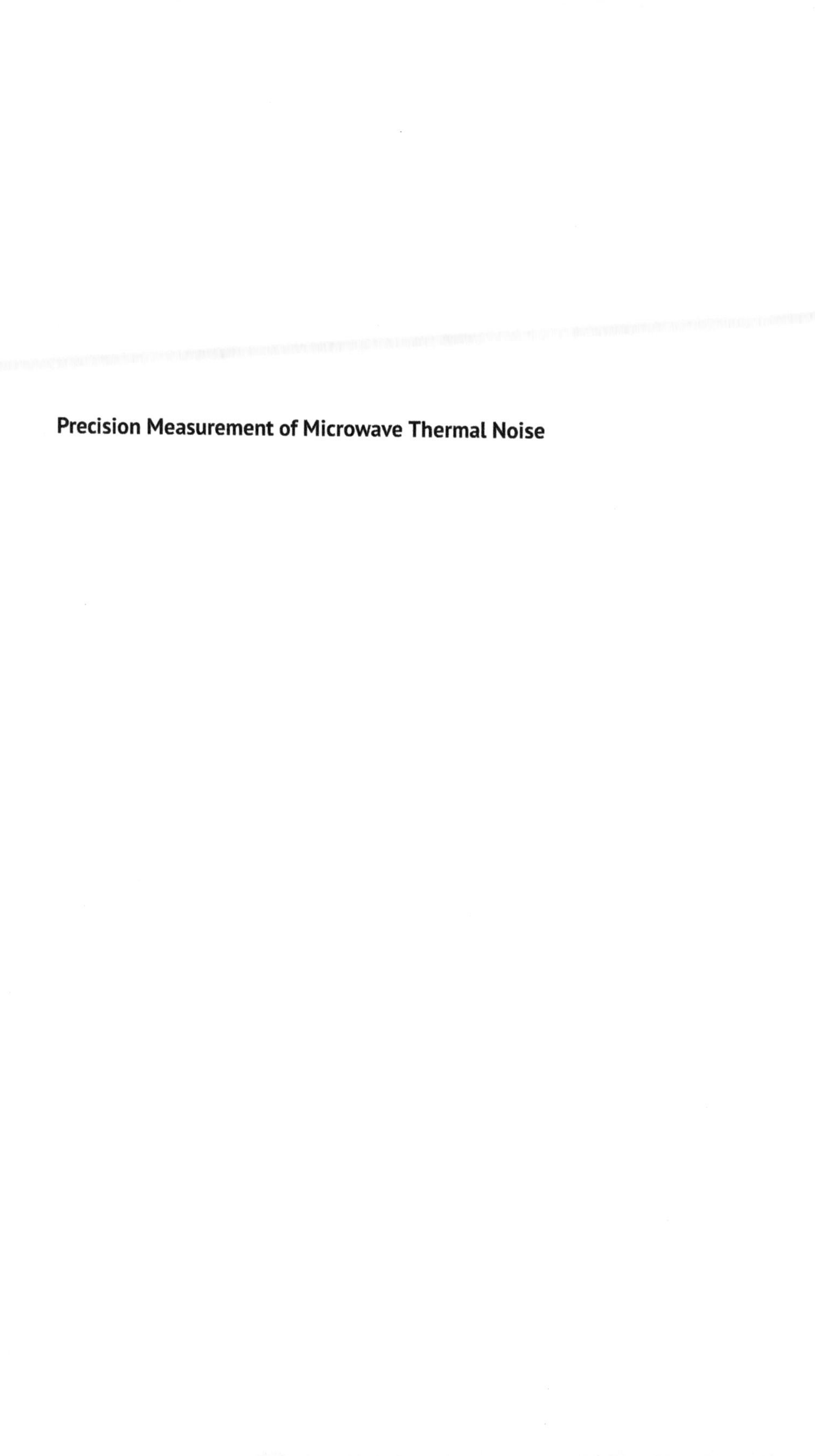

Precision Measurement of Microwave Thermal Noise

Precision Measurement of Microwave Thermal Noise

James Randa
Spectrum Technology and Research Division
National Institute of Standards and Technology
Boulder, USA

Published by John Wiley & Sons, Inc., Hoboken, New Jersey.
Published simultaneously in Canada.

For general information on our other products and services or for technical support, please contact our Customer Care Department within the United States at (800) 762-2974, outside the United States at (317) 572-3993 or fax (317) 572-4002.

Wiley also publishes its books in a variety of electronic formats. Some content that appears in print may not be available in electronic formats. For more information about Wiley products, visit our web site at www.wiley.com.

Library of Congress Cataloging-in-Publication Data is Applied for:

Hardback ISBN: 9781119910091

Cover Design: Wiley
Cover Image: © Pobytov/Getty Images

Set in 9.5/12.5pt STIXTwoText by Straive, Chennai, India

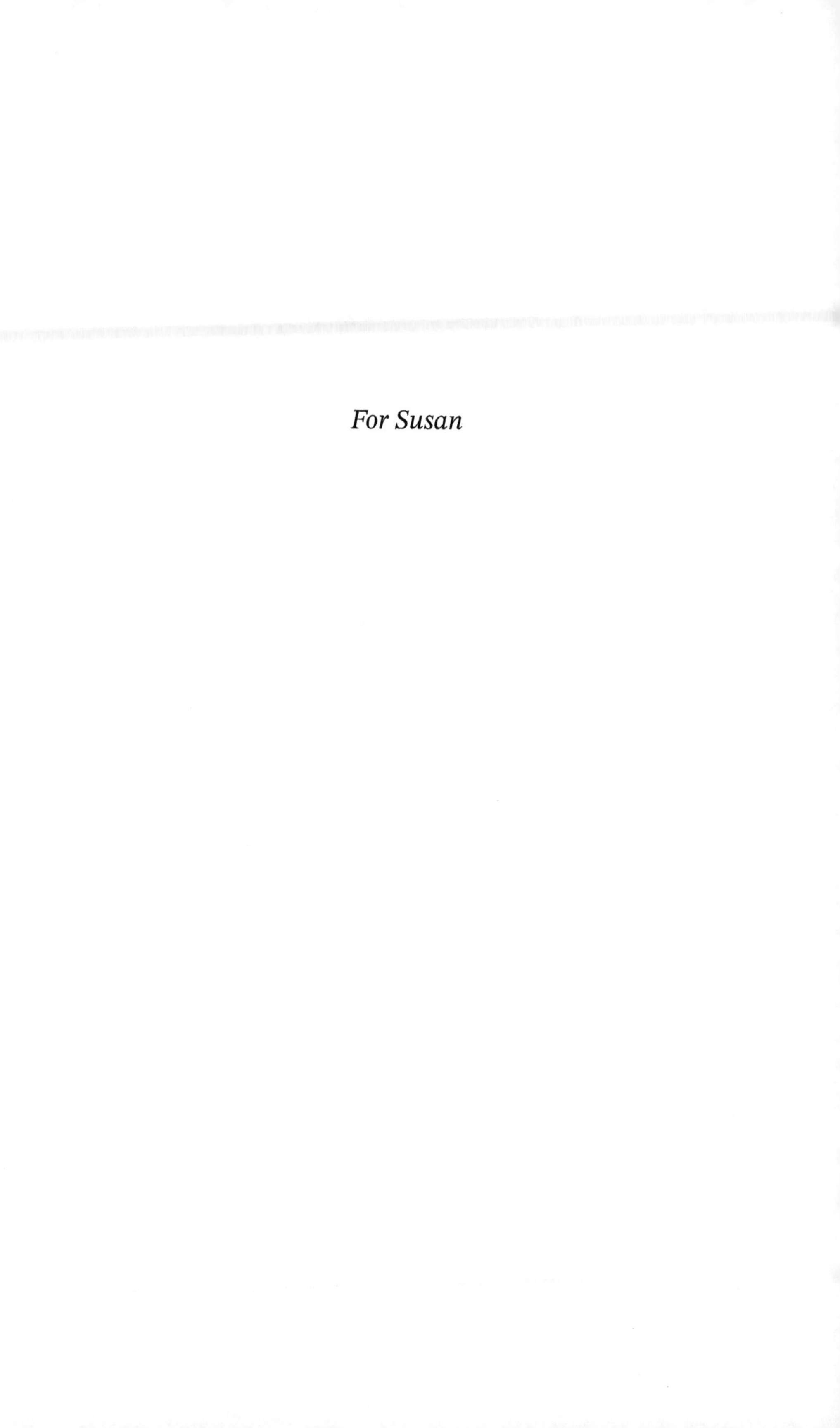

For Susan

Contents

Preface *xi*

1 Background *1*
1.1 Nyquist's Theorem and Noise Temperature *1*
1.1.1 Nyquist's Theorem *1*
1.1.2 Limits and Numbers *2*
1.1.3 Definition of Noise Temperature *4*
1.1.4 Excess Noise Ratio and T_0 *5*
1.2 Microwave Networks *5*
1.2.1 Notation *5*
1.2.2 Noise Correlation Matrix and Bosma's Theorem *6*
1.2.3 Power Ratios *7*
1.2.4 Noise-Temperature Translation Through a Passive Device *9*
References *10*

2 Noise-Temperature Standards *11*
2.1 Introduction *11*
2.2 Ambient Standards *12*
2.3 Hot (Oven) Standards *13*
2.4 Cryogenic Standards *13*
2.4.1 Coaxial Standards *13*
2.4.2 Waveguide Standards *15*
2.5 Other Standards and Noise Sources *18*
2.5.1 Tunable Primary Standards *18*
2.5.2 "Equivalent Hot Standard" Based on RF Power *18*
2.5.3 Secondary Standards *19*
2.5.4 Synthetic Primary Standards *19*
References *20*

3 Noise-Temperature Measurement *23*
3.1 Background *23*
3.2 Total-Power Radiometer *24*
3.2.1 Idealized Case *24*
3.2.2 Nonideal Case *25*
3.2.3 Radiometer Equation for Isolated Total-Power Radiometer *27*
3.2.4 Total-Power Radiometer Design *29*
3.2.5 Radiometer Testing *32*
3.3 Total-Power Radiometer Uncertainties *34*
3.3.1 Type-A Uncertainties *34*
3.3.2 Type-B Uncertainties *36*
3.3.3 Sample Results *40*
3.4 Other Radiometer Designs *40*
3.4.1 Switching or Dicke Radiometer *40*
3.4.2 Digital Radiometer *41*
3.5 Measurements through Adapters *42*
3.6 Traceability and Inter-laboratory Comparisons *43*
References *44*

4 Amplifier Noise *47*
4.1 Noise Figure, Effective Input Noise Temperature *47*
4.2 Noise-Temperature Definition Revisited *48*
4.3 Noise Figure Measurement, Simple Case *49*
4.4 Definition of Noise Parameters *50*
4.4.1 Circuit Treatment of Noisy Amplifier *50*
4.4.2 Wave Representation of Noise Parameters *52*
4.5 Measurement of Noise Parameters *55*
4.5.1 General Measurement Setup *55*
4.5.2 Fit to Noise-Figure Parameterization *59*
4.5.3 Fit to Noise-Temperature or Power Parameterization *60*
4.5.4 Possible Variations When Using the Wave Formulation *62*
4.5.5 Choice of Input Terminations *63*
4.5.6 Commercial Systems, Source-Pull Measurements *66*
4.5.7 Frequency–Variation Method *66*
4.6 Uncertainty Analysis for Noise-Parameter Measurements *67*
4.6.1 Simple Considerations *67*
4.6.2 Full Analysis *70*
4.6.3 Input Uncertainties *72*
4.6.4 General Features and Sample Results *74*
4.7 Simulations and Strategies *77*
References *79*

5 On-Wafer Noise Measurements *83*
5.1 Introduction *83*
5.2 On-Wafer Microwave Formalism *84*
5.2.1 Traveling Waves vs. Pseudo Waves *84*
5.2.2 On-Wafer Reference Planes *84*
5.3 Noise-Temperature Measurements *85*
5.4 On-Wafer Noise-Parameter Measurements *88*
5.4.1 General *88*
5.4.2 Radiometer-Based Systems *90*
5.4.3 Commercial Systems and Reference-Plane Considerations *93*
5.4.4 "Enhanced" or Model-Assisted Measurements *95*
5.5 Uncertainties *101*
5.5.1 Differences from Packaged Amplifiers *101*
5.5.2 General Features and Properties *103*
5.5.3 Measurement Strategies *104*
References *105*

6 Noise-Parameter Checks and Verification *109*
6.1 Measurement of Passive or Previously Measured Devices *109*
6.2 Physical Bounds and Model Predictions *111*
6.3 Tandem or Hybrid Measurements *112*
References *118*

7 Cryogenic Amplifiers *121*
7.1 Background *121*
7.1.1 Introduction *121*
7.1.2 Vacuum-Fluctuation Contribution *121*
7.2 Measurement of the Matched Noise Figure *123*
7.2.1 Cold-Attenuator Method *123*
7.2.2 Internal Hot–Cold Method *124*
7.2.3 Full-Characterization Measurements *125*
7.3 Noise-Parameter Measurement *128*
References *129*

8 Multiport Amplifiers *133*
8.1 Introduction *133*
8.2 Formalism and Noise Matrix *134*
8.3 Definition of Noise Figure for Multiports *136*
8.4 Degradation of Signal-to-Noise Ratio *138*
8.5 Three-Port Example – Differential Amplifier with Reflectionless Terminations *139*

8.5.1 Motivation *139*
8.5.2 Characteristic Noise Temperature, Gains, and Effective Input Noise Temperature *139*
8.5.3 Noise Figure *142*
8.5.4 Practical Applications *143*
8.6 Four-Port Example with Reflectionless Terminations *143*
References *145*

9 Remote Sensing Connection *147*
9.1 Introduction *147*
9.2 Theory for Standard Radiometer *149*
9.3 Standard-Radiometer Measurements *154*
9.3.1 Determination of α *154*
9.3.2 Determination of Illumination Efficiency, η_{IE} *154*
9.3.2.1 Measurements of a Standard Target *155*
9.4 Standard-Target Design *155*
9.5 Target Reflectivity Effects *156*
9.5.1 Effect of Target Reflectivity *156*
9.5.2 Measurement of Target Reflectivity *157*
References *157*

Index *159*

Preface

Precision noise measurements are among the most difficult measurements in the microwave realm. The difficulty stems from the miniscule powers that must be measured, and it is compounded by the necessity of accurately measuring multiple reflection coefficients and scattering parameters in order to make the requisite corrections to the power measurements. Fortunately, in many circumstances, simplifying assumptions is justified and imprecise noise measurements suffice. Even in such cases, however, it is necessary to understand and evaluate the full, unsimplified case (at least approximately) in order to determine whether the simplifying approximations are justified.

This book attempts to present the basics of precise measurements of thermal noise at microwave frequencies. The focus is on measurement methods used at the U.S. National Institute of Standards and Technology (NIST), but it is hoped that the general principles and methods will be useful in a wide range of applications. For the most part, the material will comprise methods and theoretical underpinnings, rather than details of instrumentation. In part, this reflects the author's own expertise, but it also permits an emphasis on the general procedures and overall measurement framework. If we consider a qualitative "comprehensibility spectrum" (SI units not yet defined), the target spectrum for this book would be roughly represented by this sketch.

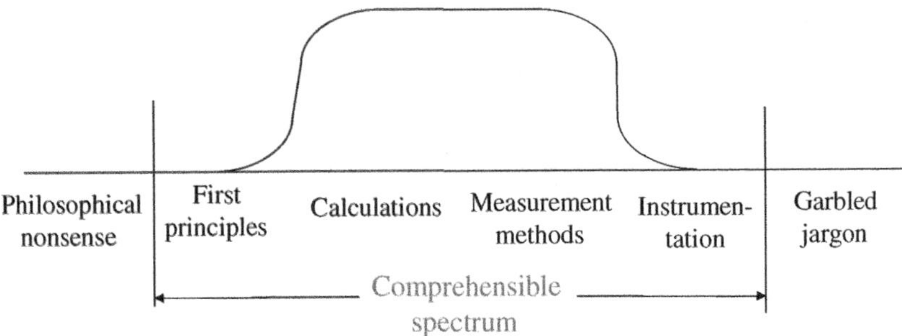

The presentation will be light on first principles and details of instrumentation, instead emphasizing calculations and measurement methods.

Just what constitutes a "precision measurement" varies with the field. Time and frequency measurements are made to 18 significant figures, whereas in other fields uncertainties may be measured in decibels. We will use the criterion that the measurements and methods of interest are those with the lowest uncertainties, such as those made at National Measurement Institutes (NMIs). For thermal noise measurements, this is typically on the order of tenths of a percent. The emphasis will be on accuracy, with only secondary consideration (if that) given to the time required for the measurement. Since "accurate" measurements are defined to be those with the smallest uncertainties, considerable attention will be paid to uncertainty analysis of the various measurement methods. Included in that discussion will be some mention of traceability, which underpins the accuracy of the underlying standards. In some cases, the highest accuracy may be attained by commercially available instrumentation and methods. In such cases, we will try to present the principles underlying the methods rather than the design and use of the instrumentation. The material covered is not groundbreaking or new; however, we hope to present a solid basis for this somewhat arcane but nevertheless important field.

Microwave radiometry originated in remote sensing of celestial bodies, and many or most major advances occurred in that field. This book, however, is concerned with laboratory noise measurements. We do not cover remote-sensing applications, but we do discuss the relevance of laboratory noise measurements to them.

Much of this book is based on previously published work by the author, referenced in the relevant sections, and from various lectures at Automatic Radio Frequency Techniques Group (ARFTG) short courses and at other laboratories. The material of Chapters 4–6 is based on two earlier review articles [1, 2].

No one works or learns in a vacuum, and I owe a debt of gratitude to numerous colleagues at NIST. These include (but are not limited to) Dave Wait and Bill

Daywitt, from whom I learned much about the foundations of the field; Dazhen Gu and Dave Walker, with whom I subsequently learned much new material and who provided fabrication and measurement expertise; and two exceptional technicians, Jack Rice and the late Rob Billinger, who provided fabrication support and precise, reliable measurement results on which any theoretical developments depended. The remote-sensing work benefitted greatly from the efforts of Amanda Cox as a post-doc and Derek Houtz as a student. To them and the many other contributors to the NIST Noise Project as well as collaborators elsewhere, thank you.

I am grateful to the editors at Wiley, first for choosing to publish this work and then for their efforts and assistance in overcoming the various hurdles that stand between a submitted manuscript and a real, material book.

On a personal note, I appreciate greatly the companionship and support of family and friends throughout my life. In particular, I am grateful to my parents, John and Catherine (Kay) Randa, for providing life lessons and a firm, stable basis from which to launch myself into the wide world. And, of course, I am grateful to my wife Susan for her continuing support, tolerance, and love through the years.

References

1 J. Randa, "Amplifier and transistor noise-parameter measurements," in *Wiley Encyclopedia of Electrical and Electronics Engineering*, Wiley; ed. J. Webster (2014). doi: 10.1002/047134608X.W8219.

2 J. Randa, "Numerical modeling and uncertainty analysis of transistor noise-parameter measurements, *The International Journal of Numerical Modelling*, Wiley Online Library (wileyonlinelibrary.com) (2014). doi: 10.1002/jnm.2039.

1

Background

1.1 Nyquist's Theorem and Noise Temperature

1.1.1 Nyquist's Theorem

Conduction electrons in a physical resistor at nonzero temperature are in continual thermal motion, and at any instant this motion induces a voltage v across the terminals of the resistor. Because the motion is random, the induced voltage averages to zero, $\langle v(t)\rangle = 0$, but the average squared voltage is nonzero, $\langle v^2(t)\rangle \neq 0$, and therefore electrical power can be extracted from the resistor. This phenomenon was first measured by Johnson [1] and was explained by Nyquist [2]. The entire field of thermal noise measurement rests on Nyquist's theorem,[1] which relates the mean square voltage across a resistor due to thermal motion of its electrons to the physical temperature of the resistor,

$$\langle v^2(f)\rangle df = 4R(f)\frac{hdf}{\left(e^{hf/k_BT} - 1\right)} \tag{1.1}$$

where $v(f)$ is the voltage in the differential frequency interval df centered at frequency f, $R(f)$ is the real part of the impedance at f, h is Planck's constant, and k_B is Boltzmann's constant.

For microwave frequencies, it is convenient to cast Eq. (1.1) in the form of an equation for the power available from the resistor,

$$\langle P_{avail}(f)\rangle = \frac{hf}{e^{hf/(k_BT)} - 1}\Delta f \tag{1.2a}$$

$$\langle p_{avail}(f)\rangle = \frac{hf}{e^{hf/(k_BT)} - 1} \tag{1.2b}$$

where $P_{avail}(f)$ is the available power in the interval Δf centered at f, and $p_{avail}(f)$ is the spectral available power density. In general, we will use upper case P to refer to

1 This is one of two fundamental Nyquist theorems that underpin entire fields (or subfields) of study, the other being his (and Shannon's) sampling theorem.

Precision Measurement of Microwave Thermal Noise, First Edition. James Randa.

power and lower-case p to refer to spectral power density. The brackets in Eqs. (1.2) indicate an ensemble or time average (assumed to be the same).

Nyquist's original derivation relied on an analysis of propagation modes in a lossless transmission line. He first derived the classical result (those were the early days of quantum mechanics), which assumed that the total energy per degree of freedom was equal to $k_B T$. That led to

$$\langle v^2(f)\rangle df \approx 4R(f)k_B T\Delta f \tag{1.3a}$$

or

$$\langle P_{avail}(f)\rangle \approx k_B T\Delta f \tag{1.3b}$$

He then noted that if instead of $k_B T$ the total energy per degree of freedom was $hf/(e^{hf/kT}-1)$ then the result was Eq. (1.1), and thus Eqs. (1.2). We have taken the liberty of using the approximately equal sign in Eqs. (1.3) in order to indicate that the true result is given by Eqs. (1.1) and (1.2).

Nyquist's treatment and result are reminiscent of the problem of black-body radiation from a heated object, and the thermal noise in an electrical circuit is in fact the one-dimensional version of black-body radiation. As in the black-body radiation case, the classical result, given by Eqs. (1.3), is plagued by the "ultraviolet catastrophe," the fact that the total power available is infinite when one integrates over all frequencies. The quantum factors provide the necessary damping at high frequency keeping the total energy available finite. A modern, full-quantum (i.e. second-quantized) treatment of Nyquist's theorem can be found in [3]. An interesting property that emerges in the full quantum treatment is that an auxiliary field (the noise field) is actually *required* for a linear two-port in order for the quantum commutation relations to be consistent (unless the $\boldsymbol{S}$ matrix is the identity matrix or the temperature is zero).

1.1.2 Limits and Numbers

It is instructive to consider the general behavior of the function in Eqs. (1.2). Figure 1.1a,b plots the available power spectral density as a function of frequency on a logarithmic and a linear scale, respectively, for different values of the physical temperature. There is a broad plateau that extends up to high frequency, where the spectral power density drops off precipitously. The low-frequency behavior is given by expanding Eqs. (1.2) for small f,

$$\langle P_{avail}\rangle \approx k_B T\Delta f\left[1-\frac{hf}{(2k_B T)}\right] \approx k_B T\Delta f \tag{1.4}$$

which is a constant, independent of frequency, depending only on the physical temperature. Furthermore, that constant is very small; even for a temperature of 10 000 K, the power density in a 1 MHz bandwidth is only 0.138 pW.

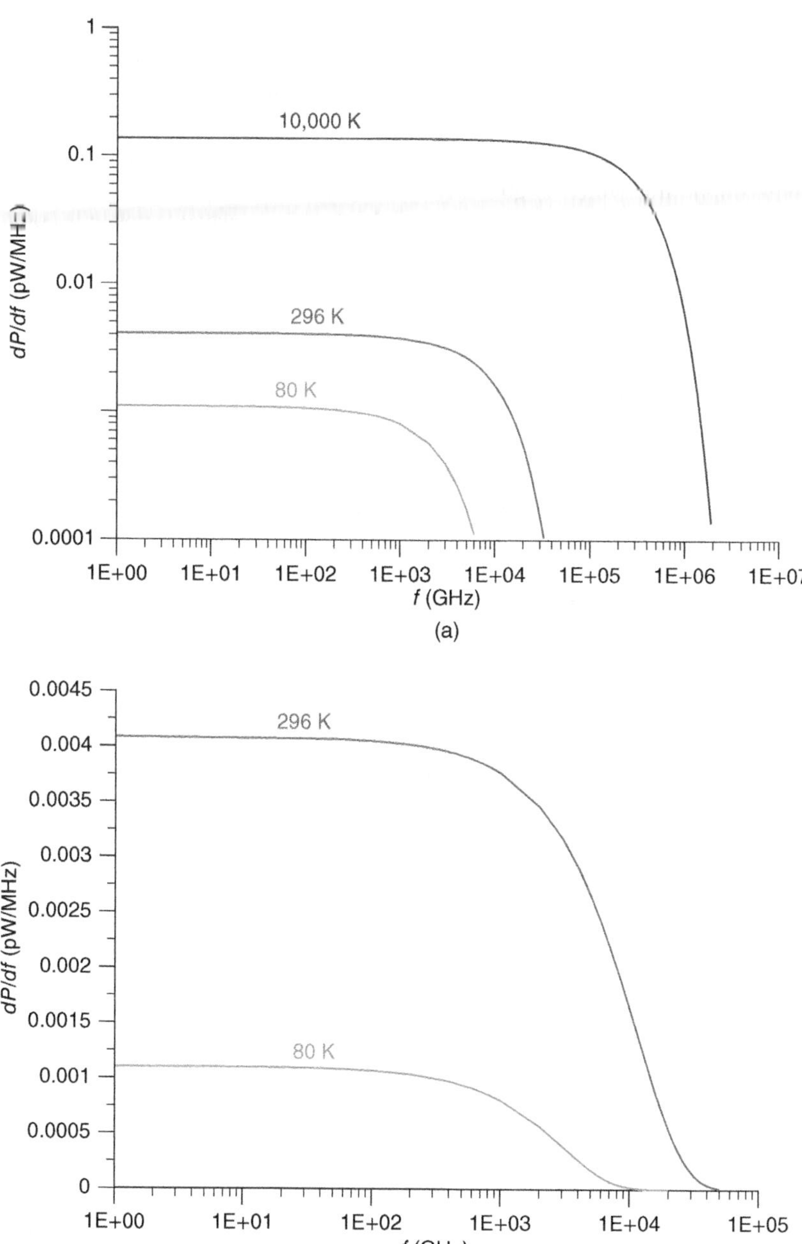

Figure 1.1 Available power spectral density as a function of frequency; (a) logarithmic scale, (b) linear scale.

The high-frequency behavior of the available power density is dominated by the exponential in the denominator, which drives the power rapidly to zero once $hf/(k_BT)$ becomes sizable. The "knee" in the graphs, where the behavior transitions from the low-frequency constant to the high-frequency damping, occurs at around $f(\text{GHz}) \approx 20\times T(\text{K})$. This transition occurs when quantum effects become important, which is governed by the value of $h/k_B = 0.04799\,\text{K/GHz}$. Thus departures from the simple constant behavior of the available power become important for very high frequency and/or very low temperature. For example, at 290 K it is a 1% effect at 116 GHz; at 100 K it is a 1% effect at 40 GHz and a 0.1% effect at 4 GHz; at 30 K and 40 GHz it is a 6.4% effect (about 0.26 dB).

1.1.3 Definition of Noise Temperature

Equations (1.2) relate the available noise power from a passive device to its physical temperature. But microwave circuitry involves more than passive components. It is therefore convenient to define a "noise temperature" for active devices. Many variations have been suggested (and used), but there are two principal ways to do this [4]. The first is to use Eq. (1.2b) and define the noise temperature to be the physical temperature of a passive device that would result in the observed available power density. We will refer to this definition as the "equivalent-physical-temperature" definition,

$$\langle p_{avail}(f)\rangle \equiv \frac{hf}{e^{hf/k_BT_{noise}}-1} \tag{1.5a}$$

The average in Eq. (1.5a) is taken over the frequency interval Δf, centered at f. This definition is popular in the remote-sensing community, where the received power is used to measure the physical temperature of the object under observation. This definition has the appealing property that for a passive object or device, the noise temperature is simply the physical temperature. Inverting Eq. (1.5a) to obtain the equations for T_{noise} yields a rather complicated expression, a point to which we shall return when considering amplifier noise measurements in Chapter 4.

The other common choice [5], which we adopt here, is to define the noise temperature as the available power spectral density divided by the Boltzmann constant times the frequency interval, which we will call the "Power Definition,"

$$\langle p_{avail}(f)\rangle \equiv k_BT_{noise} \tag{1.5b}$$

With this definition, the noise temperature is just a surrogate for the noise power spectral density, which makes this the natural choice when dealing with microwave circuits. With the power definition, the noise temperature of a passive device or object is only approximately equal to the physical temperature,

$$T_{noise} = \frac{1}{k_B}\left[\frac{hf}{e^{\frac{hf}{k_BT_{phys}}}-1}\right] \approx T_{phys} \tag{1.6}$$

The approximation of Eq. (1.6) is known as the Rayleigh–Jeans approximation.

Due to the approximate equality in Eq. (1.6), there is *usually* little difference "in every-day life" between the power definition and the equivalent-physical-temperature definition. However, in precision measurements it is not uncommon to encounter a combination of high frequency, low temperature, and high precision that requires a specific choice of definition. Since this book deals with microwave precision noise measurements, we adopt the power definition, Eq. (1.5b), for the noise temperature. The considerations of Section 1.1.2 above explain when the high-frequency, low-temperature corrections become important, and thus when the distinction between different noise-temperature definition starts to matter.

1.1.4 Excess Noise Ratio and T_0

In dealing with noise temperatures and powers of greatly differing magnitudes, it is sometimes useful to define a decibel quantity for noise temperature. The quantity that is commonly used is the Excess Noise Ratio (*ENR*), defined by

$$ENR \equiv 10log_{10}\left(\frac{T_{noise} - T_0}{T_0}\right) \tag{1.7}$$

where the reference temperature T_0 is taken to be $T_0 = 290$ K. Just to be clear, we treat T_0 as a noise-temperature constant; it is not a noise temperature that corresponds to a physical temperature by virtue of Eq. (1.6). It is just a number.

Some variants of the *ENR* definition can be found either in the literature or in casual use. Some practitioners use the power delivered to a matched (i.e. reflectionless) load in place of the available power implicit in the T_{noise} in Eq. (1.7). This has the effect of introducing a mismatch factor (see Section 1.2.3) into the first term in the numerator of Eq. (1.7). There is also the question [6] of whether the reference temperature should be 290 K, or should it be the noise temperature of a passive load at a physical temperature of 290 K. We take it to be $T_0 = 290$ K, just for simplicity. We are venturing into the hair-splitting realm here, but in precision measurements sometimes hairs must be split.

1.2 Microwave Networks

1.2.1 Notation

Noise measurements are just a type of power measurements, and in precision noise measurements it is imperative to carefully account for all sources and flows of power. Therefore, before delving into noise measurements, it is necessary to review some basic microwave network theory and to establish the conventions

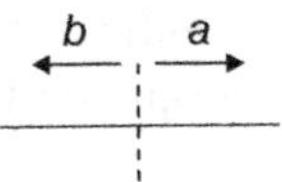

Figure 1.2 Waves on a lossless transmission line.

and notation that will be used in this work. Many books cover microwave circuit theory in some detail, including [7–12]. Here we just summarize the results that will be used. Our approach is similar to that of [13], with appropriate generalizations to include active devices. Unless specifically stated, we assume that all reference planes are in lossless transmission lines which support only a single propagation mode at the frequencies considered. We will typically use a and b to refer to the amplitudes of the traveling waves propagating in the two directions on a transmission line, with the normalization such that the spectral power density of the wave is given by the magnitude of its amplitude squared, $|a|^2$ or $|b|^2$. The net power spectral density delivered to the right in Figure 1.2 is thus given by

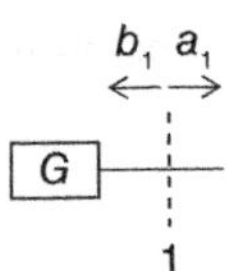

Figure 1.3 A linear one-port.

$$p_{del} = |a|^2 - |b|^2 \tag{1.8}$$

A linear one-port, as shown in Figure 1.3, is then described by

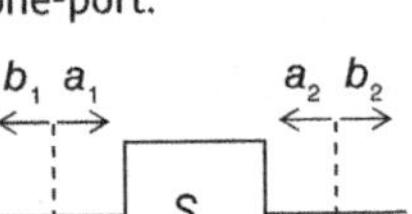

Figure 1.4 A linear two-port.

$$a_1 = \Gamma_G b_1 + c_G \tag{1.9}$$

where Γ_G is the reflection coefficient of the one-port G, and c_G is the wave emanating from G due to its intrinsic noise. A linear two-port, as depicted in Figure 1.4, is described by

$$\begin{pmatrix} b_1 \\ b_2 \end{pmatrix} = \begin{pmatrix} S_{11} & S_{12} \\ S_{21} & S_{22} \end{pmatrix} \begin{pmatrix} a_1 \\ a_2 \end{pmatrix} + \begin{pmatrix} c_1 \\ c_2 \end{pmatrix} \tag{1.10}$$

where c_1 and c_2 are the waves due to the intrinsic noise sources in the two-port.

1.2.2 Noise Correlation Matrix and Bosma's Theorem

In terms of the wave amplitudes introduced above, the noise correlation matrix of a two-port device is defined as

$$\boldsymbol{N} \equiv \begin{pmatrix} \langle |b_1|^2 \rangle & \langle b_1 b_2^* \rangle \\ \langle b_1^* b_2 \rangle & \langle |b_2|^2 \rangle \end{pmatrix} \tag{1.11}$$

This describes the noise properties of the device in a circuit, where there will in general be incident noise waves $\boldsymbol{a}$. If we are interested in the intrinsic properties of the device itself, we want to know the noise correlation matrix for the case in which there are no incident waves. That is given by the *intrinsic noise correlation matrix*,

$$\hat{\boldsymbol{N}} \equiv \begin{pmatrix} \langle |c_1|^2 \rangle & \langle c_1 c_2^* \rangle \\ \langle c_1^* c_2 \rangle & \langle |c_2|^2 \rangle \end{pmatrix} \tag{1.12}$$

In this book, we will be interested in properties of devices themselves, rather than in their use in circuits, and so we will deal primarily with the intrinsic noise matrix of Eq. (1.12). The exception will occur in Chapter 8, where we analyze multiport amplifiers.

An important result for the intrinsic noise matrix of a passive device is Bosma's theorem [12]. This states that for a passive device at noise temperature T_a the intrinsic noise-correlation matrix is related to the scattering matrix $\boldsymbol{S}$ by the equation

$$\hat{\boldsymbol{N}} = k_B T_a(\boldsymbol{I} - \boldsymbol{S}\boldsymbol{S}^{\dagger}) \tag{1.13}$$

where $\boldsymbol{I}$ is the identity matrix. This means that the noise properties of a passive device are entirely determined by its $\boldsymbol{S}$ matrix (and its temperature). Although it is used mostly for two-ports, Eq. (1.13) applies for any number of ports. For a passive two port, the elements of the intrinsic noise-correlation matrix are given explicitly by

$$\begin{aligned}
\left\langle |c_1|^2 \right\rangle &= k_B T_a \left(1 - \left[S_{11}{}^2\right] - |S_{12}|^2\right)\\
\left\langle c_1 c_2^* \right\rangle &= -k_B T_a \left(S_{11} S_{21}^* + S_{12} S_{22}^*\right)\\
\left\langle c_2 c_1^* \right\rangle &= -k_B T_a \left(S_{21} S_{11}^* + S_{22} S_{12}^*\right)\\
\left\langle |c_2|^2 \right\rangle &= k_B T_a \left(1 - \left[S_{22}{}^2\right] - |S_{21}|^2\right)
\end{aligned} \tag{1.14}$$

1.2.3 Power Ratios

The power delivered to a load L, Figure 1.5, is given by $|a_1|^2 - |b_1|^2$. If we use $b_1 = \Gamma_L a_1$, we can write

$$p_{1,del} = |a_1|^2 - |b_1|^2 = |a_1|^2 \left(1 - |\Gamma_L|^2\right) \tag{1.15}$$

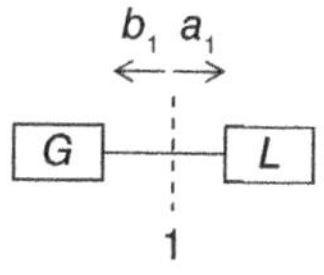

Figure 1.5 Power delivered to a load.

For the *available* power from a source G, we refer to Eq. (1.9) and Figure 1.5 and write the expression for the power delivered from the source to the load L,

$$a_1 = \Gamma_G b_1 + c_G = \Gamma_G \Gamma_L a_1 + c_G = \frac{c_G}{1 - \Gamma_G \Gamma_L}$$

$$p_{1,del} = |a_1|^2 - |b_1|^2 = \left(1 - |\Gamma_L|^2\right)|a_1|^2 = \frac{|c_G|^2}{|1 - \Gamma_G \Gamma_L|^2}\left(1 - |\Gamma_L|^2\right) \tag{1.16}$$

This delivered power has a maximum for $\Gamma_L = \Gamma_G^*$, whence

$$p_{1,avail} = \frac{|c_G|^2}{\left(1 - |\Gamma_G|^2\right)} \tag{1.17}$$

Referring to the definition of noise temperature in Eq. (1.5b), we obtain

$$\left\langle |c_G|^2 \right\rangle = \left(1 - |\Gamma_G|^2\right) k_B T_G \tag{1.18}$$

This relates the magnitude of the intrinsic noise wave to the noise temperature of the one-port device, and due to our choice of noise-temperature definition, it is exactly true for both active and passive devices.

In tracking the power flow in a microwave circuit, three ubiquitous quantities are the mismatch factor, the efficiency, and the available power ratios. The mismatch factor is the fraction of the available power that is actually delivered across a reference plane. Referring to Figure 1.5, the mismatch factor at plane 1 is defined as $M_1 \equiv p_{1,\,del}/p_{1,\,avail}$. From Eqs. (1.16) and (1.17) for the delivered and available powers respectively, we have

$$M_1 \equiv \frac{p_{1,del}}{p_{1,avail}} = \frac{\left(1 - |\Gamma_L|^2\right)\left(1 - |\Gamma_G|^2\right)}{|1 - \Gamma_L \Gamma_G|^2} \tag{1.19}$$

The efficiency is a property of a two-port that measures the fraction of the power that is delivered to the two-port which is delivered at the output of the device. It is also referred to as the delivered gain. Referring to Figure 1.6, $\eta_{21} \equiv p_{2,\,del}/p_{1,\,del}$. In terms of the relevant scattering parameters and reflection coefficients, we can write (after some algebra)

$$\eta_{21} \equiv \frac{p_{2,del}}{p_{1,del}} = \frac{|S_{21}|^2\left(1 - |\Gamma_L|^2\right)}{|1 - \Gamma_L S_{22}|^2\left(1 - |\Gamma_{SL}|^2\right)} \tag{1.20}$$

where Γ_{SL} is the reflection coefficient at plane 1 from the S–L combination. If we substitute the explicit expression for Γ_{SL} into Eq. (1.21), we obtain

$$\eta_{21} = \frac{|S_{21}|^2\left(1 - |\Gamma_L|^2\right)}{|1 - \Gamma_L S_{22}|^2 - |(S_{12}S_{21} - S_{11}S_{22})\Gamma_L + S_{11}|^2} \tag{1.21}$$

Note that the efficiency depends only on the S parameters and the reflection coefficient of the load; it is independent of the source.

Besides the efficiency, another important quantity when dealing with a two-port is the available power ratio (or available gain) α_{21}. If we refer to Figure 1.6, α_{21} is defined as the ratio of the available power at plane 2 to the available power at plane 1, disregarding any intrinsic noise sources in the load or the two-port, $c_L = c_{1,\,S} = c_{2,\,S} = 0$,

$$\alpha_{21} \equiv \frac{p_{2,avail}}{p_{1,avail}}(c_L = c_{1\cdot S} = c_{2,S} = 0) \tag{1.22}$$

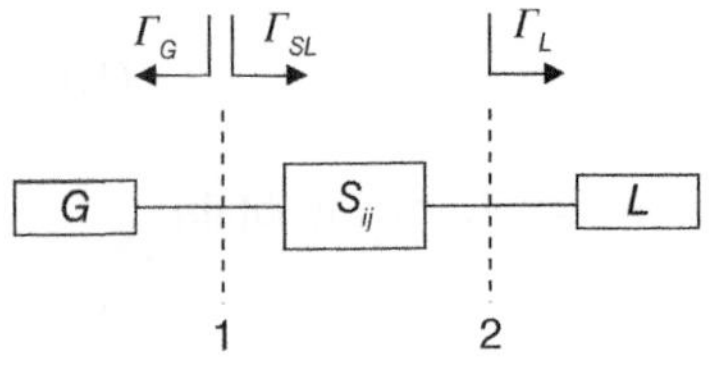

Figure 1.6 A two-port between source *G* and load *L*.

where $c_{1,S}$ and $c_{2,S}$ are the intrinsic sources of the two-port as defined by Eq. (1.10). In terms of the reflection coefficients and S parameters, Eq. (1.22) can be written as

$$\alpha_{21} = \frac{|S_{21}|^2 \left(1 - |\Gamma_G|^2\right)}{|1 - \Gamma_G S_{11}|^2 \left(1 - |\Gamma_{GS}|^2\right)} \tag{1.23}$$

where Γ_{GS} is the reflection coefficient from the G–S combination at plane 2,

$$\Gamma_{GS} = S_{22} + \frac{S_{21} S_{12} \Gamma_G}{1 - \Gamma_G S_{11}} \tag{1.24}$$

In Figure 1.6 (and in subsequent figures) the arrows indicate that the reflection coefficient is looking into and reflected from the direction of the arrow.

Whereas the efficiency was independent of the source reflection coefficient, the available power ratio is independent of the reflection coefficient of the load. Note that the available power ratio is also the *available gain*.

1.2.4 Noise-Temperature Translation Through a Passive Device

A situation that often occurs in analyzing noise measurements is depicted in Figure 1.7. The noise temperature is known at reference plane 1, and one needs to know the noise temperature at reference plane 2, on the other side of a passive device at noise temperature T_a. (Note that we specify a *noise* temperature of T_a, not a physical temperature of T_a, cf. Eq. (1.6).) The device can be any passive, linear two-port, such as an attenuator, adaptor, or section of lossy line. Because the noise temperature is defined in terms of the available power, Eq. (1.5b), it is possible to compute T_2 in terms of T_1 and T_a from the available power ratio α of the passive device. The fact that it is a linear device, plus the definition of the available power ratio mean that we can write

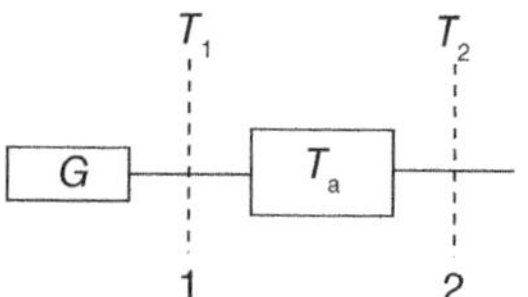

Figure 1.7 Configuration for computing T_2, knowing T_1.

$$p_{2,avail} = \alpha_{21} p_{1,avail} + f_0(T_a) \tag{1.25}$$

where f_0 is some function of T_a but does not depend on $p_{2,avail}$. From Eq. (1.5b) it then follows that

$$T_2 = \alpha_{21} T_1 + f(T_a) \tag{1.26}$$

where $f_0 = k_B f$. When $T_1 = T_a$, we must have $T_2 = T_a$, and therefore

$$T_a = \alpha_{21} T_a + f(T_a), \quad f(T_a) = (1 - \alpha_{21}) T_a \tag{1.27}$$

Combining Eqs. (1.27) and (1.28) yields the desired relationship

$$T_2 = \alpha_{21} T_1 + (1 - \alpha_{21}) T_a \tag{1.28}$$

Equation (1.28) is a fundamental result that is used extensively in analyzing noise properties of microwave circuits.

References

1 J. Johnson, "Thermal agitation of electricity in conductors," *Physical Review,* **32**, pp. 97–109 (1928). doi: 10.1103/physrev.32.97.

2 H. Nyquist, "Thermal agitation of electric charge in conductors," *Physical Review,* **32**, pp. 110–113 (1928). doi: 10.1103/physrev.32.110.

3 J. Zmuidzinas, "Thermal noise and correlations in photon detection," *Applied Optics*, **42**, pp. 4989–5008 (2003).

4 A.R. Kerr and J. Randa, "Thermal noise and noise measurements – a 2010 update," *IEEE Microwave Magazine*, **11**, no. 6, pp. 40–52 (October 2010). doi: 10.1109/MMM.2010.937732.

5 J. Randa, J. Lahtinen, A. Camps, A.J. Gasiewski, M. Hallikainen, D.M. Le Vine, M. Martin-Neira, J. Piepmeier, P.W. Rosenkranz, C.S. Ruf, J. Shiue, and N. Skou, "Recommended terminology for microwave radiometry," NIST Technical Note 1551 (August 2008). doi: 10.6028/NIST.TN.1551

6 A.R. Kerr, "Suggestions for revised definitions of noise quantities, including quantum effects," *IEEE Transactions on Microwave Theory Techniques*, **47**, no. 3, pp. 325–329 (March 1999).

7 D.M. Kerns and R.W. Beatty, *Basic Theory of Waveguide Junctions and Introductory Microwave Network Analysis*, Pergamon Press, London, 1974.

8 S. Ramo, J.R. Whinnery, and T. Van Duzer, *Fields and Waves in Communications Electronics*, Wiley, New York, 1984.

9 G.F. Engen, *Microwave Circuit Theory*, Peter Peregrinus Ltd., London, UK, 1992.

10 R.E. Collin, *Foundations for Microwave Engineering*, McGraw-Hill, New York, 1992.

11 D.M. Pozar, *Microwave Engineering*, 4th edn., Wiley, New York, 2011.

12 H. Bosma, "On the theory of linear noise systems," *Philips Research Reports Supplements*, no. 10, p. 191 (1967).

13 D.F. Wait, "Thermal noise from a passive linear multiport," *IEEE Transactions on Microwave Theory and Techniques*, **16**, no. 9, pp. 687–691 (September 1968).

2

Noise-Temperature Standards

2.1 Introduction

Measurements of noise temperature rely on comparison to a noise-temperature standard, a source of noise power with a known noise temperature. The standard's noise temperature may be calculable from first principles, in which case it is a *primary* standard, or it may be known by comparison to some other standard, in which case it is referred to as a *secondary* standard. (In principle, one can define *tertiary*, etc. standards in the obvious way, but for our present purposes that's a hair we need not split.) An important usage of secondary standards is as *transfer* standards, used in place of a primary standard (with a resulting increase in uncertainty) for reasons of convenience or accessibility. We will be most concerned with primary standards since they are more accurate than secondary standards. Technically, a secondary standard based on an extremely accurate primary standard could be more accurate than an inferior primary standard, but again we choose not to wander into such semantic morasses.

Most primary noise standards are based on Nyquist's theorem, discussed in Chapter 1. A resistance is maintained at a known physical temperature, and its noise temperature is given by

$$T_{noise} = \frac{1}{k_B}\left[\frac{hf}{e^{hf/(k_B T_{phys})} - 1}\right] \approx T_{phys} \tag{2.1}$$

where again the approximation is the Rayleigh–Jeans approximation. Since the entire point of a primary noise standard is to have an accurately known noise temperature, the approximation is not used in precision measurements. The known quantity in such a thermal primary standard is the *available* spectral power density, so in principle the reflection coefficient of the standard is not important, as long as it is known. In practice, however, the uncertainty in the standard is smaller if the standard is matched (reflectionless) or nearly so. The known physical temperature is usually achieved by surrounding the load

Precision Measurement of Microwave Thermal Noise, First Edition. James Randa.

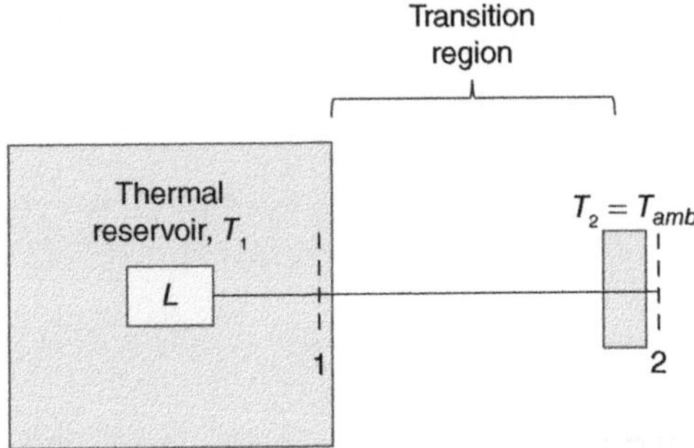

Figure 2.1 Depiction of elements of a primary noise standard.

with some medium at a known, constant temperature. The temperature of the surrounding medium can be known by direct measurement, or it can be based on some physical property such as the boiling temperature of liquid nitrogen (LN).

The general design of a standard based on Nyquist's theorem is outlined in Figure 2.1. A resistive load is surrounded by and in thermal contact with a heated or cooled medium maintained at a constant physical temperature T_1, resulting in the load itself being at physical temperature T_1. The noise temperature of the load is thus given by Eq. (2.1). The load is connected to the outside world through a transition region extending from plane 1 to plane 2, which is at the ambient temperature. The transition region is necessary if the standard is to be used with or connected to equipment in a normal laboratory environment at a temperature different from T_1. Typically, it is the characterization of this transition transmission line that is a principal source of uncertainty in the standard's noise temperature. This is because any losses in the transition line result in the addition or subtraction of noise power (depending on whether T_1 is less or greater than T_{amb}) and from Eq. (1.28) the amount of noise added depends on the amount of loss and the physical temperature of the line segment in which it occurs. The transition-region correction thus depends on the distribution of loss and of physical temperature along the transition transmission line.

If the standard is to be connected to a device or system at the same physical temperature, then one can dispense with the transition region, and lower uncertainties can be achieved. This is the case with ambient-temperature standards, which are considered in Section 2.2. We will see another example of this when we consider measurements on cryogenic amplifiers in Chapter 7.

2.2 Ambient Standards

The simplest and most accurate noise standards are those designed to operate at or near the ambient laboratory temperature. Most noise-temperature measurements require at least two different standards, and an ambient standard is virtually always employed as one of the two, due to its convenience and accuracy. An ambient standard can be just a resistive load left to equilibrate to room temperature, but for precision measurements it is best to stabilize and insulate the load from potential fluctuations in the ambient temperature. At the U.S. National Institute of Standards and Technology (NIST), this is typically done by circulating water from a temperature-controlled bath so that the load is in thermal contact with the tubing carrying the circulating water and insulated from the surrounding air [1]. In this manner, the uncertainty in the noise temperature of the ambient standard can be kept to about 0.2 K, which is small enough to be nearly negligible compared to other uncertainties.

An additional caution regarding the "ambient" temperature: it is sometimes used to refer to the (noise) temperature of the ambient standard and sometimes used to refer to the room temperature, which are not necessarily the same. It is sometimes necessary to distinguish between the temperature of the ambient standard and the ambient or room temperature (as well as between the physical temperature and the noise temperature of course).

2.3 Hot (Oven) Standards

The most accurate primary noise standards tend to be hot standards based on some sort of oven technology [2–5]. These were also some of the earliest primary noise standards. For such standards, the temperature T_1 in Figure 2.1 is well above ambient, ranging up to roughly 1000 K or more.

Because of the difficulty of using oven standards, they are seldom used in day-to-day calibration measurements. The common practice is to use them to calibrate secondary standards, which are then used in routine measurements, with a consequent increase in the uncertainty.

2.4 Cryogenic Standards

2.4.1 Coaxial Standards

Although they are usually not quite as accurate as oven standards, cryogenic primary noise standards [6, 7] are popular because they are more robust and easier to use than typical oven standards. Because of their ease of use, they can be used in routine, everyday measurements and calibrations – unlike hot standards, which typically employ a transfer standard for everyday use.

The reason that cryogenic primary standards tend to have somewhat higher uncertainties is that fractional errors in the correction for the transition section are larger for cold standards than for hot standards. This is because the noise temperature that the transition section mixes with the reservoir temperature T_1 is larger than T_1 for a cold standard and less than T_1 for a hot standard, and thus any error in how much of the transition temperature is added or subtracted is a larger effect relative to T_1 for a cold T_1 than for a hot T_1. Nonetheless, cryogenic primary standards can be very accurate and are usually considerably more accurate than secondary standards calibrated with a hot primary standard.

The design of a coaxial cryogenic standard is shown in Figure 2.2. The design was done for NIST by W. Daywitt; the figure is from [8]. A matched load is immersed in a bath of LN and connected to the output connector by a coaxial transmission line (PC-7 in this case). The LN is contained in a Dewar to minimize heat transfer from the surroundings to the LN. The container is not sealed so that the surface of the LN is exposed to atmospheric pressure and is at its boiling temperature. The load is located at a fixed location in the container and is at

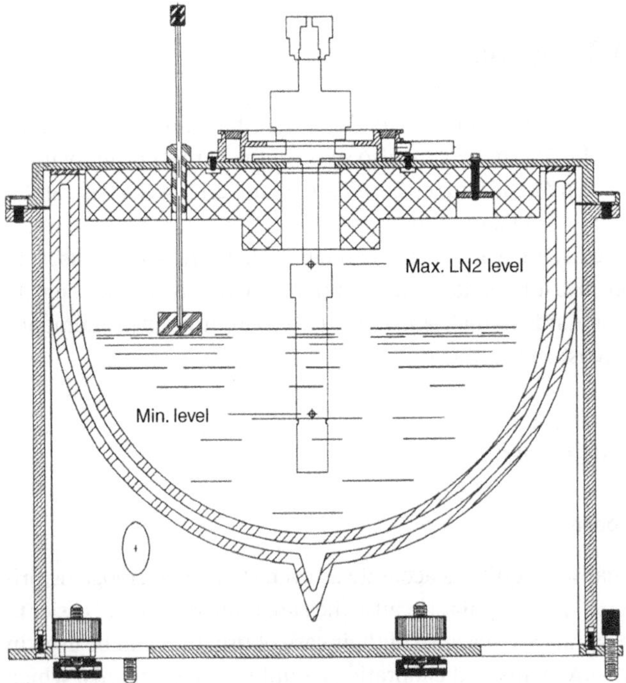

Figure 2.2 Sketch of the design of a coaxial cryogenic standard. Source: Designed for NIST by W. Daywitt. Figure from [8].

some operating depth within the LN. A float is used to indicate the height of the top of the LN bath, and it is monitored to keep the depth of the load within the operating range. If the LN surface falls too low, the standard is recharged with additional LN. (More modern commercial models monitor and control the depth automatically.) The physical temperature of the load is corrected for the variation of the temperature with depth within the LN.

The surface temperature of the LN is the boiling point of LN, corrected for atmospheric pressure, which is measured by a manometer in the laboratory. Since the standard was developed and is used at NIST in Boulder, CO, at about 1646 m (5400 ft) above sea level, deviations from sea-level atmospheric pressure can be significant. The temperature of the load is taken to be that of the surrounding LN, which is calculated for the operating depth below the LN surface.

The output connector is maintained at room temperature by circulating water from a constant-temperature water bath. Considerable care was taken in the design of the transmission line connecting the load to the output connector. The outer conductor is in thermal contact with the LN at the bottom and the output connector at the top, and thus its temperature is known at those points, and is assumed to vary linearly in the region between. To keep the inner conductor at the same temperature as the outer conductor, the beads used as spacers to support the inner conductor need to have high thermal conductivity. On the other hand, the spacers must also be electrical insulators in order to preserve the electromagnetic properties of the transmission line. And they must have low dielectric coefficients in order to minimize reflections within the length of the line. Few materials meet all these qualifications. One that does is the ceramic material beryllium oxide, which unfortunately is a nightmare to machine. Nonetheless, it was used to fabricate the beads. The transmission line, load, output section, and inner conductor with beads are shown in Figure 2.3.

The PC-7 standard is only used up to 12.4 GHz, since above that frequency better uncertainties are obtained using the WR-62 waveguide standard and system. Over the frequency range where it is used, the operating noise temperature of the PC-7 standard is between 80 K and 85.5 K, depending on frequency, with a standard fractional uncertainty of 0.63–0.77% in the output noise temperature. The higher temperatures and uncertainties occur at the higher frequencies due to increased resistive losses.

2.4.2 Waveguide Standards

The design of a WR-10 waveguide cryogenic standard [7, 9, 10] is shown in Figure 2.4, and photos of a disassembled waveguide standard are shown in Figure 2.5. The load consists of silicon carbide wedges inside a copper cylinder, surrounded by LN. The copper cylinder is attached to a stainless-steel cylinder

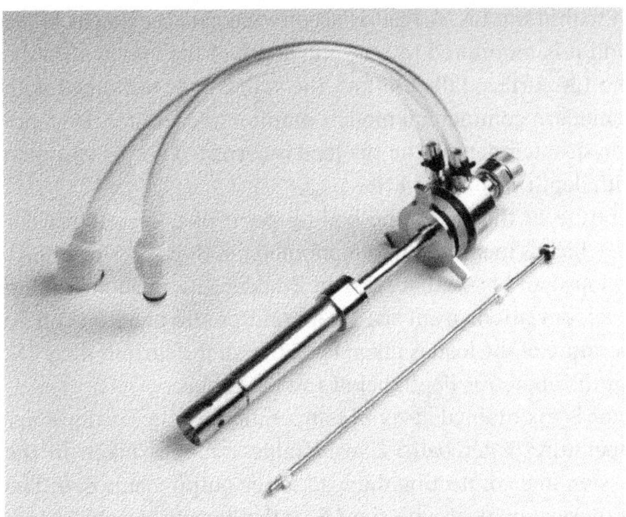

Figure 2.3 Load, transmission line, and output used in coaxial cryogenic standard, showing inner conductor with beads. Source: J. Randa [8]/W. Daywitt./National Institute of Standards and Technology.

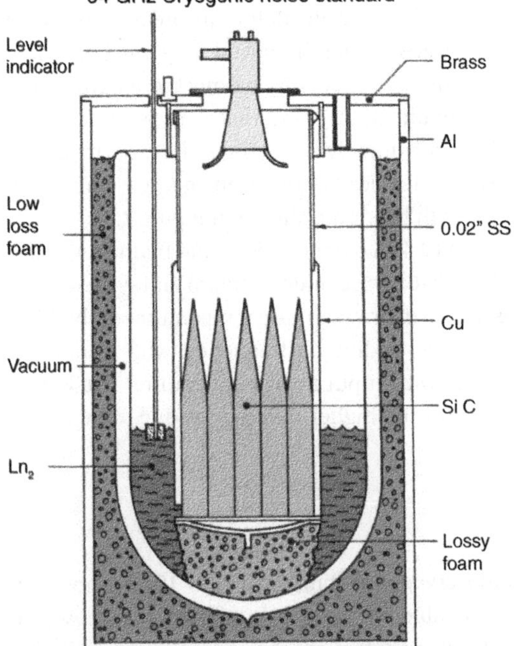

Figure 2.4 Design of a waveguide cryogenic standard [7]. Source: Figure taken from Randa [8].

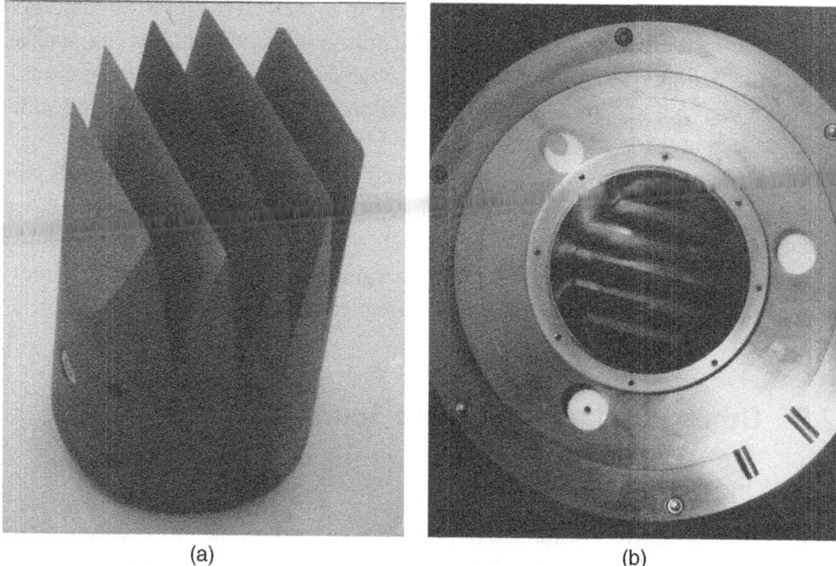

Figure 2.5 a Silicon carbide wedges used in cryogenic waveguide standard; b Wedges inside containing cylinder, viewed from above. Cylinder diameter is about 13 cm. Source: Randa [8]/National Institute of Standards and Technology.

above it, to limit the heat flow from the outer structure on which the output horn sits. Only the base of the wedges is immersed in the LN, which is then wicked up and boils off the wedge surfaces, maintaining them at the boiling temperature of the LN. That the wedge surfaces are at the boiling temperature of LN was verified by direct measurement.

The transition region in these waveguide standards consists of the copper and stainless-steel cylinders above the LN. The horn at the output is flared to increase the gain and reduce the power received from the cylinder walls. The temperature of the horn is maintained at ambient temperature by means of circulating water from a constant-temperature water bath. The operating noise temperatures of the waveguide cryogenic standards range from about 76 to 78 K in the frequency range 12.4–60 GHz (WR-62 to WR-15), with standard (1-sigma) fractional uncertainties ranging from 0.17% to 0.48%. The operating temperatures and uncertainties vary from band to band due to resistive losses and differences in the fabrication of the various standards.

The waveguide standards generally have slightly lower uncertainties than the coaxial standards at the same frequency due to the relative difficulty of characterizing the losses and reflections in the coaxial line as opposed to the antenna–absorber configuration.

At NIST, coaxial systems are usually used up to 12.4 GHz and waveguide systems are used at higher frequencies. Coaxial noise sources above 12.4 GHz, such as PC-7, PC-3.5, and PC-2.4, are measured through adapters on waveguide radiometers with waveguide standards. (Only precision connectors are considered sufficiently repeatable for good calibrations.)

Cryogenic primary standards are now used at national primary standards laboratories besides NIST [11–13], and various versions of cryogenic primary noise standards are available commercially as well. Some of the commercial standards include features such as automatically recharging the LN and availability of interchangeable waveguide horns for different frequency bands [14].

2.5 Other Standards and Noise Sources

2.5.1 Tunable Primary Standards

At low frequencies (below about 0.8–1 GHz), isolators become prohibitively large. As we shall discuss in Chapter 3, this requires either that a radiometer have its noise properties fully characterized as a function of the impedance (reflection coefficient) of the input source or that it be calibrated with standards having the same reflection coefficient as the device to be measured. For this purpose, NIST developed tunable primary standards for use at 30 and 60 MHz [15]. Because the tuning capability is restricted to a narrow range of frequencies, these standards can each be used only at its design frequency.

2.5.2 "Equivalent Hot Standard" Based on RF Power

Because noise measurements are measurements of power, it is natural to consider relating noise standards to existing microwave power standards. Two obstacles impede a direct link to power standards, but they can be overcome or circumvented. The first is that noise temperature is a power spectral density, not a power, and so the measurement system bandwidth is needed if we are to relate the two. The other difficulty is that microwave power measurements and standards are made at vastly larger power levels than noise measurements. A noise temperature of 1000 K corresponds to a power spectral density of about 1.38×10^{-11} mW/MHz, whereas microwave power measurements or calibrations are typically done at powers of a few or several mW.

The Swiss Federal Institute of Metrology (METAS) has developed an "equivalent hot standard" for noise temperature based on microwave power standards [16]. By measuring the equivalent noise bandwidth of their microwave power measurement system and employing a sizeable attenuator to reduce the hot noise temperature to a desirable level, they have constructed a noise-temperature standard based on their power standard. It has not yet been compared fully to other noise-temperature standards, but initial indications are positive.

2.5.3 Secondary Standards

Secondary noise standards are those whose noise temperatures are known by comparison to primary standards by means of a noise-source calibration. They are typically used as transfer standards, to transfer the known noise temperature of the primary standard to a situation or application for which the primary standard is unavailable or inconvenient to use. The key features of a good secondary standard are that it be stable and repeatable, easy to use, robust, and have noise temperature and reflection coefficient that are suitable for the intended purpose. Some of the earliest secondary standards were gas discharge tubes [17], but by far the most common commercially available secondary standards are now solid-state noise sources [18].

The typical solid-state noise source consists of an avalanche diode followed by output circuitry. The output circuitry serves three purposes. Because the avalanche diode usually has a bare noise temperature much greater than what is desired, the output circuitry usually includes a significant attenuator to reduce the output noise temperature to the desired range. This attenuator also serves the second purpose of providing a better match to a 50 Ω transmission line. Finally, the output circuitry may include conditioning elements to provide an output frequency spectrum that better approximates white noise. Diode noise sources of this sort are available commercially from several sources and at various noise temperatures. The most popular noise temperatures are around 1200 K and near 9500 K. They span a wide frequency range and include most popular connector types and waveguide sizes. They usually are well matched and quite stable and repeatable over long time periods [19].

Although most secondary standards are "hot," with operating noise temperature well above ambient, it is also possible to construct cold transfer standards [20–22]. These can be made by terminating the output of a low-noise amplifier with a matched load and using the amplifier input as the noise source. Although not widely available commercially, a basic cold noise source is not particularly difficult to fabricate, and we shall see in Chapters 4 and 5 that they can prove useful in certain measurements on low-noise amplifiers and transistors.

2.5.4 Synthetic Primary Standards

An appealing possibility for a primary noise standard is to digitally synthesize the noise signal [23] using a quantum-based digital waveform synthesizer. Such a standard would be traceable to fundamental constants via the Josephson voltage. Until recently, the frequency range of the technology has been too low to be applicable for microwave noise applications, but recently the upper operating frequency of Josephson arbitrary waveform synthesizers (JAWS) has been extended to 1 GHz [24]. Since such systems operate at cryogenic temperatures, there would still be the problem of characterizing and accounting for the transition from cryogenic temperature to room temperature, but if a noise-temperature standard could be developed based on this technology, it would be very interesting to compare such synthetic standards to the current physical standards.

References

1 C.A. Grosvenor, J. Randa, and R.L. Billinger, "Design and testing of NFRad—A new noise measurement system," NIST Technical Note TN1518 (March 2000). https://tsapps.nist.gov/publication/get_pdf.cfm?pub_id=6264 (Accessed 17 March, 2022).

2 W.C. Daywitt, "A reference noise standard for millimeter waves," *IEEE Transactions on Microwave Theory and Techniques*, **MTT-21**, no 12, pp.845–847 (December 1973).

3 M.W. Sinclair and A.M. Wallace, "A new national electrical noise standard in X-band," *Proceedings of IEE*, **133**, pt. A, no. 5, pp. 272–274 (July 1986).

4 J. Achkar, "A set of waveguide primary thermal noise standards and related calibration systems for the frequency range 8.2–40 GHz," *IEEE Transactions on Instrumentation and Measurement*, **48**, no. 2, pp. 638–641 (April 1999).

5 F. Buchholz and W. Kessel, "A new primary thermal noise standard at PTB for the frequency range 12.4–18.0 GHz," *IEEE Transactions on Instrumentation and Measurement*, **42**, no. 2, pp. 258–263 (April 1993). doi: 10.1109/19.278561.

6 W.C. Daywitt, "A coaxial noise standard for the 1 GH to 12.4 GHz frequency range," National Bureau of Standards (USA), NBS Technical Note 1074 (1984).

7 W.C. Daywitt, "Design and error analysis for the WR10 thermal noise standard," National Bureau of Standards (USA), NBS Techical Note 1071 (December 1983).

8 J. Randa, "Short introduction to noise measurements at NIST," Lecture delivered at Beijing Institute of Radio Metrology and Measurement (BIRMM), Beijing (11 October 2012).

9 W.C. Daywitt, "A derivation for the noise temperature of a horn-type noise standard," *Metrologia*, **21**, no. 3, pp. 127–133 (1985).

10 W.C. Daywitt, "The noise temperature of an arbitrarily shaped microwave cavity with application to a set of millimetre wave primary standards," *Metrologia*, **30**, no. 5, pp. 471–478 (1994).

11 T. Kang, J. Kim, N. Kang and J. Kang, "A thermal noise measurement system for noise temperature standards in W-band," *IEEE Transactions on Instrumentation and Measurement*, **64**, no. 6, pp. 1741–1747 (June 2015). doi: 10.1109/TIM.2015.2398957.

12 T.-W. Kang, J.H. Kim, E.F. Yurchuk, J.-I. Park, M.V. Sargsyan, I.E. Arsaev, and R.I. Ouzdin, "Design, construction, and performance evaluation of a cryogenic 7-mm coaxial noise standard," *IEEE Transactions on Instrumentation and Measurement*, **56**, no. 2, pp. 439–443 (April 2007).

13 A. Díaz-Morcillo, A. Lozano-Guerrero, J. Fornet-Ruiz, and J. Monzó-Cabrera, "Analysis of noise temperature sensitivity for the design of a broadband thermal noise primary standard," *Metrologia,* **49**, pp. 538–551 (2012). doi: 10.1088/0026-1394/49/4/538

14 D.R. Vizard, P.R. Foster, B. Lunn and S.M. Cherry, "Millimetre wave primary noise standards in waveguide bands 18–170 GHz," *2007 European Microwave Conference*, pp. 352–355 (2007). doi: 10.1109/EUMC.2007.4405199.

15 C.A. Grosvenor and R.L. Billinger, "The NIST 30/60 MHz tuned radiometer for noise temperature measurements," NIST Technical Note TN1525 (May 2002).

16 D. Stalder, M. Wollensack, J. Hoffmann, and M. Zeier, "Traceable noise temperature calibration based on RF power," *Talk given at European Microwave Week* 2019, Porte de Versailles Paris, France, 29 September to 4 October, 2019.

17 M.W. Mumford, "A broadband microwave noise source," *The Bell System Technical Journal*, **28**, pp. 608–618 (October 1949).

18 W.T. Read, "A proposed high-frequency, negative resistance diode," *The Bell System Technical Journal*, **37**, pp. 401–406 (1958).

19 J. Randa, L.P. Dunleavy, and L.A. Terrell, "Stability measurements on noise sources," *IEEE Transactions on Instrumentation and Measurement*, **50**, no. 2, pp. 368–372 (April 2001).

20 R.H. Frater and D.R. Williams, "An active 'cold' noise source," *IEEE Transactions on Microwave Theory and Techniques,* **29**, pp. 344–347 (1981).

21 L.P. Dunleavy, M.C. Smith, S.M. Lardizabal, A. Fejzuli, and R.S. Roeder, "Design and characterization of FET based cold/hot noise sources," in *Digest 1997 IEEE MTT-S International Microwave Symposium*, Denver, CO, June 1997, pp. 1293–1296.

22 M.H. Weatherspoon and L.P. Dunleavy, "Experimental validation of generalized equations for FET cold noise source design," *IEEE Transactions on Microwave Theory and Techniques*, **54**, no. 2, pp. 608–614 (February 2006).

23 S.W. Nam, S.W. Nam, S.P. Benz, P.D. Dresselhaus, W.L. Tew, D.R. White, and J.M. Martinis, "Johnson noise thermometry measurements using a quantized voltage noise source for calibrations," *IEEE Transactions on Instrumentation and Measurement*, **52**, no. 2, pp. 550–554 (April 2003).

24 P.F. Hopkins, J.A. Brevik, M. Castellanos-Beltran, C.A. Donnelly, N.E. Flowers-Jacobs, A.E. Fox, D. Olaya, P.D. Dresselhaus, and S.P. Benz, "RF waveform synthesizers with quantum-based voltage accuracy for communications metrology," *IEEE Transactions on Applied Superconductivity,* **29**, no. 5, Art. No. 1301105 (August 2019).

3

Noise-Temperature Measurement

3.1 Background

As the name implies, a microwave radiometer measures radiated microwave power. Originally such instruments were used to measure the power received by microwave antennas. Here we consider radiometers used in a lab environment, where the noise power is "radiated" by a source into a transmission line that conducts it to the radiometer. The transmission line can be either coaxial or waveguide. The relation to remote-sensing applications will be discussed in Chapter 9.

Since we will be discussing measurements in a "lab environment," we should specify what that environment is. The most critical property for noise measurements is that the lab and the equipment in it be maintained at a constant, known temperature. This is necessary for stability and repeatability, both of the measuring equipment and of the device under tests (DUTs). The properties of active elements, such as amplifiers or diode noise sources, can be quite temperature dependent. Passive elements with appreciable loss, such as isolators, also add a temperature-dependent amount of noise to a measurement, cf. Eq. (1.28).

Maintaining a constant laboratory temperature is not necessarily sufficient to ensure that the temperatures of active or lossy components remain constant, however. Due to the poor thermal conductivity of air, their temperature can be raised by the heat that they themselves generate. Consequently, it is often necessary to take further precautions to guarantee stability. For this reason, NIST mounts active and lossy elements of radiometers on "water plates," with good thermal contact between component and water plate. These are copper or aluminum plates in which channels are machined, and water from a temperature-controlled water bath is circulated through copper tubing fitted in the channels.

The temperature of the noise lab at NIST is maintained at $23\,^{\circ}\mathrm{C} \pm 0.5\,^{\circ}\mathrm{C}$. Relative humidity is also controlled and is set at $40\% \pm 5\%$. Very high or low relative humidity tends to have deleterious effects on electronic equipment.

Precision Measurement of Microwave Thermal Noise, First Edition. James Randa.

3.2 Total-Power Radiometer

3.2.1 Idealized Case

We will delve into the design and workings of the total-power radiometer in some detail, both because it is a popular choice for laboratory applications and because it is the configuration with which we are most familiar, since it is used at NIST [1, 2]. The workings of a total-power radiometer can be understood by referring to the simplified overview of Figure 3.1.

The input to a linear receiver is switched alternately among three noise sources: two standard sources and the DUT. In this context, the hot standard is merely the standard with the higher noise temperature; it is not necessarily hotter than ambient. The hot standard could be ambient temperature if the cold standard is a cryogenic standard; or the hot standard could be hotter than ambient, and the cold standard could be ambient.

Consider first the simplest case, the matched, symmetrical case, in which all transmission lines and switches are lossless, reflection coefficients are zero, and all switch paths are equal. The fact that all reflection coefficients vanish means that all delivered powers are equal to the available powers, and the fact that the receiver is linear implies that the reading of the power meter P_{out} is related to the power delivered to the receiver by

$$P_{out} = a + bp_{del} = a + bp_{avail} \tag{3.1}$$

where a and b are the two calibration constants required to calibrate a linear system, and where the bandwidth of the power measurement has been absorbed into the constant b. Because of our definition of the noise temperature in terms of the available power, Eq. (1.5b), p_{avail} could be replaced by $k_B T_{in}$ in Eq. (3.1). The calibration constants can be determined by measuring the output power for two different known input noise temperatures, the hot and cold standards,

$$\begin{aligned} P_h &= a + bk_B T_h \\ P_c &= a + bk_B T_c \end{aligned} \tag{3.2}$$

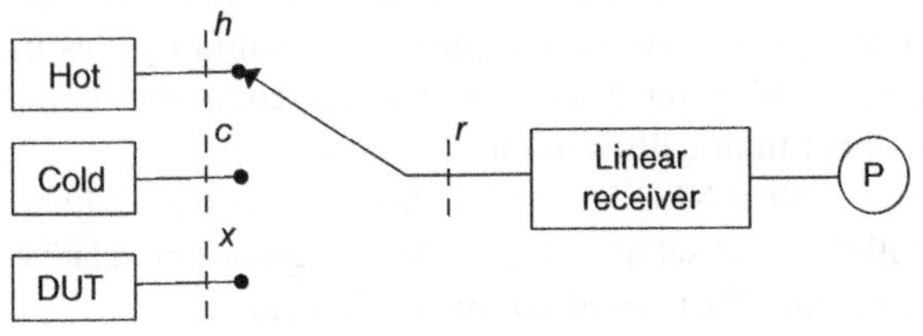

Figure 3.1 Overview of a total-power radiometer.

Solving for a and b yields

$$a = P_c - bk_B T_c, b = \frac{P_h - P_c}{k_B(T_h - T_c)} \tag{3.3}$$

If we then write Eq. (3.1) for the DUT (indicated by a subscript x), substitute Eq. (3.3) for a and b, and solve for the noise temperature of the DUT (T_x), we get

$$T_x = T_c + \frac{(Y_x - 1)}{(Y_h - 1)}(T_h - T_c) \tag{3.4}$$

where $Y_x = P_x/P_c$ and $Y_h = P_h/P_c$. Equation (3.4) is referred to as the "radiometer equation" for this radiometer; it gives the noise temperature of the DUT in terms of the noise temperatures of the two standards and the measured powers from the two standards and the DUT.

As a side note, Eq. (3.2) could also be written as

$$T_{out,Rec} = G_{rec}(T_{in} + T_{rec}) \tag{3.5}$$

where $T_{out,rec}$ is the output noise temperature from the receiver, G_{rec} is the available gain of the receiver, T_{in} is the noise temperature attached to the receiver input, and T_{rec} is the noise temperature added by the receiver. In this formulation, the gain and receiver noise temperature are given by

$$G = \frac{T_{out,h} - T_{out,c}}{T_h - T_c} \tag{3.6}$$

$$T_{rec} = \frac{Y(T_h - T_c)}{(Y - 1)} - T_h \tag{3.7}$$

where $y = T_{out,h}/T_{out,c}$. This formulation will be revisited in Chapter 4, where we consider noise in amplifiers in detail.

3.2.2 Nonideal Case

In the real world, transmission lines and switches are not perfectly lossless, switch ports are not all created equal, and reflections are nonzero. These imperfections introduce three complications into Eq. (3.1) and therefore into the analysis leading to the radiometer equation. (There is also the fact that real receivers are not perfectly linear, but we will address this issue when we consider the uncertainties.) The three complications are: (i) because the noise sources (standards and DUT) are not perfectly matched (reflectionless), the delivered power from each source is not equal to its available power (noise temperature), (ii) the power delivered by the noise source at plane h, c, or x is not the same as the power delivered to the radiometer at plane r, and (iii) the constants a and b, which relate the output power reading of the receiver to the power delivered to the input, now depend on the reflection coefficient of the input source, $a = a(\Gamma)$, $b = b(\Gamma)$.

The first two complications are handled by measuring and correcting. To do so, we must be a bit less cavalier about available vs. delivered power and just where the power is available or delivered. The relevant reference planes indicated in Figure 3.1 are at the output of the hot standard (reference plane h), at the output of the cold standard (c), at the output of the DUT (x), and at the input to the receiver or radiometer (r).

The first complication is that because the noise sources (standards and DUT) are not perfectly matched (reflectionless), the delivered power from each source is not equal to its available power (noise temperature). Since we are interested in available powers (noise temperatures) but can only measure delivered powers, we need to insert the conversion between the two, which is given by the mismatch factor of Eq. (1.19). Thus $p_{h,del} = M_h p_{h,avail}$, $p_{c,del} = M_c p_{c,avail}$, and $p_{x,del} = M_x p_{x,avail}$. The mismatch factors are given by

$$M_s = \frac{\left(1 - |\Gamma_s|^2\right)\left(1 - |\Gamma_{r,s}|^2\right)}{|1 - \Gamma_s \Gamma_{r,s}|^2} \tag{3.8}$$

where Γ_s is the reflection coefficient of the noise source s, $\Gamma_{r,s}$ is the reflection coefficient at plane s from the direction of the radiometer, and s can be h, c, or x.

The second complication is that the power delivered by the noise source at plane h, c, or x is not the same as the power delivered to the radiometer at plane r. The two are related by the efficiency factor, Eq. (1.20) or (1.21), between the two planes,

$$\begin{aligned} p_{del,r} &= \eta_{rs} p_{del,s} \\ \eta_{rs} &= \frac{|S_{21}|^2\left(1 - |\Gamma_r|^2\right)}{|1 - \Gamma_r S_{22}|^2\left(1 - |\Gamma_{r,s}|^2\right)} \end{aligned} \tag{3.9}$$

where Γ_r is the reflection coefficient of the radiometer at plane r, where the scattering matrix is from plane s to plane r, and where s can be h, c, or x, as above.

The third complication is that constants a and b now depend on the reflection coefficient of the input source, $a = a(\Gamma)$, $b = b(\Gamma)$. This property will be dealt with in Chapter 4, when we deal with noise in amplifiers. One way of understanding it is that the receiver emits noise from its input directed back toward the input noise source. Some of that noise is reflected back into the receiver, and the amount of this reflection depends on the reflection coefficient of the input noise source.

There are three ways to deal with this complication. One way is to develop tunable noise standards, whose reflection coefficients can be tuned to match that of the DUT [3]. Then the constants that characterize the receiver are the same for all the input noise sources. The second way is to insert an isolator in the input path, immediately before or after plane r in Figure 3.1 [4]. This isolates the receiver from the effects of the input noise source reflection coefficient. The third solution is to perform detailed measurements on the receiver and characterize its behavior as a function of the input reflection coefficient [5], measuring a and b (or equivalently the receiver gain and noise temperature) as functions of Γ.

At NIST, the preferred approach is to insert an isolator in front of the receiver. This is probably the easiest solution to the problem, but it has the disadvantage that isolators have limited bandwidth (compared with what could be achieved by the rest of the system). Consequently, at NIST, a separate radiometer has been built for each isolator band. An alternative approach would be to build a bank of isolators covering different bands and switch among the isolators [6]. Measuring frequency dependence of the radiometer characteristics [5] allows a wider bandwidth for the radiometer. However, it is a more labor-intensive approach, and it introduces additional sources of uncertainty. Developing tunable standards combines the disadvantages of both the other approaches. It is more work, and the resulting standards tend to have a limited frequency range. It has been used at NIST at low frequencies [3], where isolators are not available.

3.2.3 Radiometer Equation for Isolated Total-Power Radiometer

Dealing with the real-world complications leads to changes in the radiometer equation (3.4). If an isolator is inserted at plane r in Figure 3.1, we obtain Figure 3.2. If we assume an ambient standard is chosen as one of the standards and denote the other standard by S, then the equations for the powers delivered at plane 0 for each of the input sources (x, S, amb) are

$$\begin{aligned} p_{0,x}^{del} &= M_{0,x}\alpha_{03}p_x^{avail} + M_{0,x}(1-\alpha_{03})k_B T_{amb} + p_{rec,x}^{del} \\ p_{0,S}^{del} &= M_{0,S}\alpha_{02}p_S^{avail} + M_{0,S}(1-\alpha_{02})k_B T_{amb} + p_{rec,S}^{del} \\ p_{0,amb}^{del} &= M_{0,amb}\alpha_{01}p_{amb}^{avail} + M_{0,amb}(1-\alpha_{01})k_B T_{amb} + p_{rec,amb}^{del} \\ &= M_{0,amb}p_{amb}^{avail} + p_{rec,amb}^{del} \end{aligned} \tag{3.10}$$

where $M_{0,x}$ is the mismatch factor at plane 0 when the source x is connected, $p_{rec,x}^{del}$ is the effective input power to account for the receiver noise contribution when x is connected, and so on. The reading on the power meter in Figure 3.1 will be just the gain times the delivered power at plane 0. Because of the isolator, that gain will be independent of the noise source that is connected.

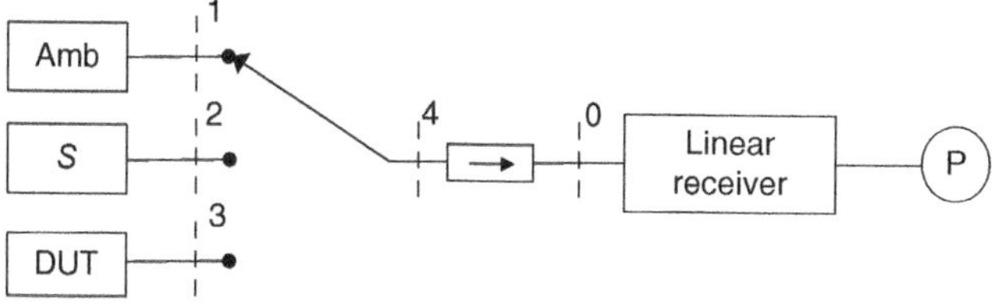

Figure 3.2 Reference planes for an isolated total-power radiometer.

We can then use $M_{0,x}\alpha_{03} = M_3\eta_{03}$, $M_{0,S}\alpha_{02} = M_S\eta_{02}$,

$$\begin{aligned}
p_{0,x}^{del} &= M_3\eta_{03}\left(p_x^{avail} - p_{amb}^{avail}\right) + M_{0,x}p_{amb}^{avail} + p_{rec,x}^{del}\\
p_{0,S}^{del} &= M_3\eta_{02}\left(p_S^{avail} - p_{amb}^{avail}\right) + M_{0,S}p_{amb}^{avail} + p_{rec,S}^{del}\\
p_{0,amb}^{del} &= M_{0,amb}p_{amb}^{avail} + p_{rec,amb}^{del}
\end{aligned} \tag{3.11}$$

Due to the isolator, $M_{0,x} = M_{0,S} = M_{0,amb}$ and $p_{rec,x}^{del} = p_{rec,S}^{del} = p_{rec,amb}^{del}$, so that Eq. (3.11) can be written as

$$\begin{aligned}
p_{0,x}^{del} &= M_3\eta_{03}\left(p_x^{avail} - p_{amb}^{avail}\right) + p_{rec,amb}^{del}\\
p_{0,S}^{del} &= M_S\eta_{02}\left(p_S^{avail} - p_{amb}^{avail}\right) + p_{rec,amb}^{del}\\
p_{0,amb}^{del} &= M_{0,amb}p_{amb}^{avail} + p_{rec,amb}^{del}
\end{aligned} \tag{3.12}$$

Any effects of imperfect isolation are included in the uncertainty [4, 7, 8]. If we subtract $p_{0,amb}^{del}$ from $p_{0,x}^{del}$ and $p_{0,S}^{del}$, and divide one by the other, we get

$$\frac{p_{0,x}^{del} - p_{0,amb}^{del}}{p_{0,S}^{del} - p_{0,amb}^{del}} = \frac{M_x\eta_{03}\left(p_x^{avail} - p_{amb}^{avail}\right)}{M_S\eta_{02}\left(p_S^{avail} - p_{amb}^{avail}\right)} \tag{3.13}$$

The power measured by the radiometer for each of the three noise sources (S, x, amb) is just a constant times the delivered power at plane 0 for each source, $P_x^{meas} = GBp_{0,x}^{del}$, and similarly for S and x, where G is an overall system gain and B is the effective system bandwidth. If we define the measured power ratios $Y_x \equiv P_x^{meas}/P_{amb}^{meas}$ and $Y_S \equiv P_S^{meas}/P_{amb}^{meas}$, and use $p^{avail} = k_BT$, we can solve Eq. (3.13) for T_x, with the result

$$T_x = T_{amb} + \frac{M_S\eta_{02}(Y_x - 1)}{M_x\eta_{03}(Y_S - 1)}(T_S - T_{amb}) \tag{3.14}$$

Equation (3.14) is the radiometer equation for an isolated total-power radiometer. We see that it differs from the idealized case of Eq. (3.4) by the presence of the ratios of mismatch factors and efficiencies. In most practical cases, neglect of these two ratios results in an error of around 1% or 2%.

Equation (3.14) is very robust in that it is essentially (as in Eq. (3.13)) a ratio of two differences. Thus, additive errors in the power measurements are removed in the differences, and multiplicative errors in mismatches and efficiencies or in power measurements are removed in the ratios. Any small residual effects of the source reflection coefficients depend on the degree of isolation; they can be dealt with in the uncertainty analysis.

If reference plane 0 is not easily accessible, Eq. (3.14) can be rewritten using the observation that $\eta_{02}/\eta_{03} = \eta_{42}/\eta_{43}$, and thus one can measure the ratio of efficiencies to the input of the isolator.

3.2.4 Total-Power Radiometer Design

A block diagram of a total-power radiometer is shown in Figure 3.3, and a photo of the NIST coaxial radiometer is shown in Figure 3.4. Figures 3.3 and 3.4 are of the NIST coaxial radiometer NFRad [1]. Referring to Figure 3.3, the front-end switch alternates among the three different input sources, the ambient standard, the cryogenic standard, and the DUT. In practice, the NFRad switch head can accommodate multiple DUTs. It allows up to seven input noise sources. Since two of the input sources must be the two standards, up to five can be DUTs. One of these DUT ports is usually used for a check standard. The check standard is a noise source that has been previously measured (multiple times); it is measured as a check, to ensure that the system is functioning as it has in the past.

Considerable care was taken with the design and fabrication of the switch head, which houses the ambient standard as well as the switch and connecting lines to the input and output ports of the switch. A photo of the disassembled switch head is shown in Figure 3.5. It is fabricated of stainless steel and has channels machined in it to carry circulating water from a constant-temperature water bath. The connectors for the input and output water are at the bottom of the switch head, at the right in Figure 3.6. The small cylinder with multiple white wires emerging is the ambient standard. The wires connect to a platinum resistance thermometer within

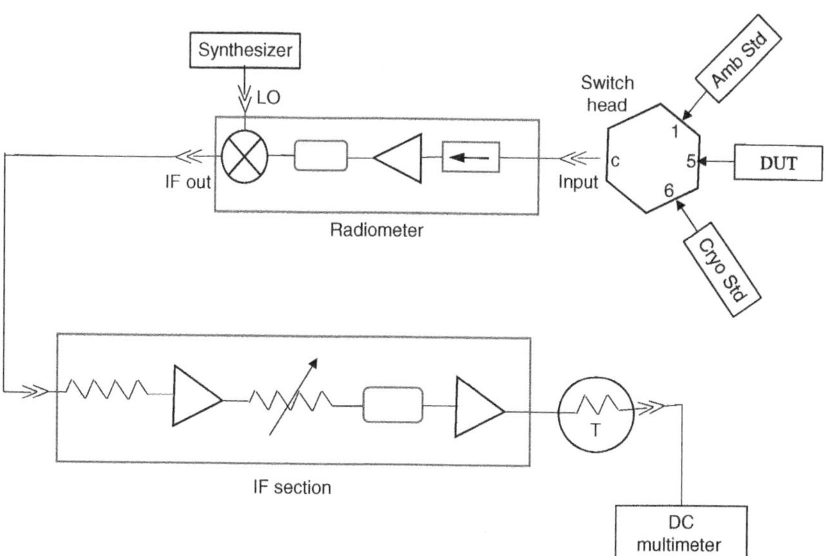

Figure 3.3 Block diagram of a total-power radiometer. Source: Grosvenor et al. [1]/U.S. Department of Commerce/Public Domian.

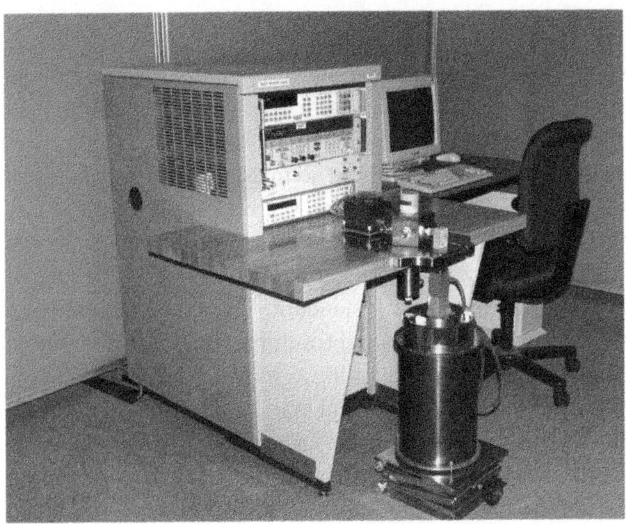

Figure 3.4 Photo of the NIST coaxial radiometer NFRad. Source: Grosvenor et al. [1]/National Institute of Standards and Technology.

Figure 3.5 Photo of the disassembled switch head of the NIST coaxial radiometer NFRad. Source: Grosvenor et al. [1]/National Institute of Standards and Technology.

the cylinder to monitor the physical temperature of the ambient standard. The gasket separates the input and output channels of the circulating water. The actual switch is mounted on top of the switch head. In the photo, it is behind the left-hand hexagonal housing, with the terminals protruding into the interior of the housing.

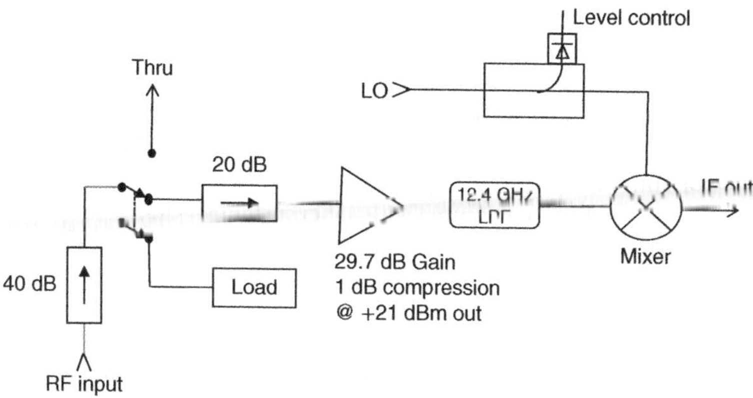

Figure 3.6 Details of receiver circuit. Source: Grosvenor et al. [1]/U.S. Department of Commerce/Public Domian.

The entire housing provides a stable controlled temperature at the input to the receiver.

The common port of the switch is connected to the input of the radiometer receiver. The receiver circuit is depicted in Figure 3.6. The radiometer input starts with an isolator, to isolate the radiometer noise characteristics from the effects of the source reflection coefficient, as discussed in Section 3.2.3. The noise power density emerging from the isolator is the isolator efficiency η_{isol} times the input power density plus the noise power added by the isolator, $\eta_{isol}p_{in} + p_{isol}$, where p_{in} is the input power density to the receiver. Following the isolator is a low-noise amplifier, which amplifies the output of the isolator and adds some noise of its own, resulting in a power density of $G(\eta_{isol}p_{in} + p_{isol} + k_B T_e)$ at the output of the amplifier, where G is the gain and $Gk_B T_e$ is the noise power density added by the amplifier. This noise added by the RF amplifier is essentially what determines the sensitivity of the radiometer, and therefore it is important that it be a low-noise amplifier.

Following the low-noise RF amplifier is a low-pass filter (LPF, 5 MHz for the coaxial system) that prevents higher-frequency components from reaching the mixer input and being down-converted into the measurement band. The mixer is double balanced to reduce harmonic mixing. An automatic level control is used on the local oscillator (LO) to maintain constant LO input to the mixer. Any noise from the LO only changes the gain and effective input noise temperature of the system. (The simplest way to see this is to expand the output signal in a power series in the RF signal. Because the RF signal is small, only the lowest two terms contribute, and these contribute to the effective input noise temperature and gain of the receiver. One can get the same result using a circuit analysis with a simple model for the mixer.) The LO is set to the nominal measurement frequency, and down-conversion occurs at baseband with both sidebands used.

The intermediate frequency (IF) section consists of two high-gain amplifiers, some matching circuitry, and some variable attenuation that is used to test linearity. Details can be found in [1]. The noise figures of the IF amplifiers are not critically important, since the desired signal has already been amplified by the RF amplifier. The output power is measured with a thermistor detector.

It is important that all active or lossy components are temperature stabilized. Properties of active components can be temperature dependent, and the derivation of the radiometer equation assumes that the receiver does not change over the course of the measurements of the different input sources. Similarly, the amount of noise added by lossy components depends on the physical temperature of those components, cf. Eq. (1.28). Consequently, in the NIST radiometer NFRad all active and lossy components are either mounted on water plates or surrounded by tubing carrying circulating water (where all the water is from a temperature-controlled bath).

3.2.5 Radiometer Testing

Before measurements on a new radiometer can be trusted, it is necessary to test the performance of the radiometer and to verify that it is operating as it should. An overall operational test of the system is to measure a DUT whose noise temperature is known and trusted (if such a DUT is available), for example one that was measured on earlier radiometers or possibly at another trusted laboratory. While this is a good test of the overall system, other more focused tests must also be used to check specific assumptions and particular aspects of the radiometer's performance.

The derivation of the radiometer equation (3.14), assumes an isolated radiometer, and therefore the isolators must be measured to verify their performance. The stability of the system must be checked, in order to assure that its properties do not change or drift over the course of the series of measurements (DUT, ambient standard, and non-ambient standard) that are required for a noise-temperature measurement. The primary standards must be checked.

Linearity of the system is critical, particularly when a large extrapolation is required from primary standard noise temperatures to the DUT noise temperature. For example, if cryogenic and ambient primary standards are used to measure a DUT noise temperature of about 8500 K, the extrapolation looks like Figure 3.7. In this case, the measured power from the DUT is 6 mW, as indicated by the horizontal long-dash line. The two calibration points are the two stars near the origin, and the linear extrapolation, shown as the solid line, results in a DUT noise temperature of about 8550 K, as indicated by the left-most short-dash

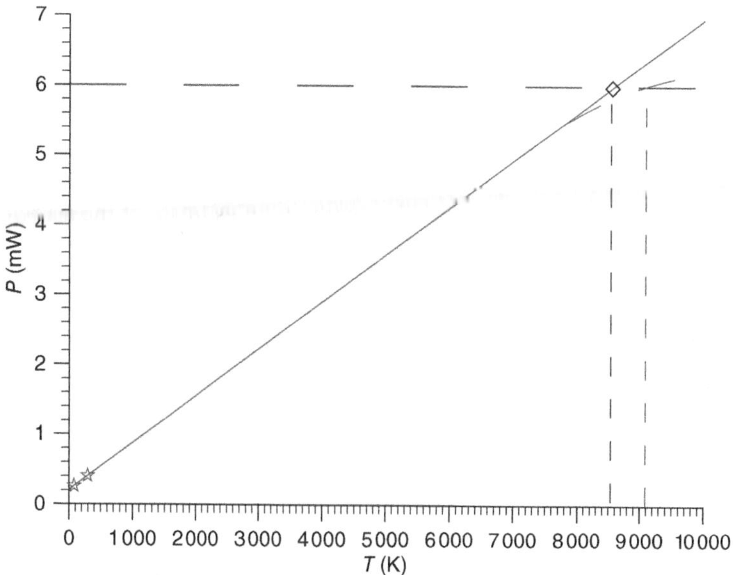

Figure 3.7 Extrapolation involved when measuring hot noise source with cryogenic and ambient primary standards.

vertical line. If, however, there was some small nonlinearity present for the higher power levels, such as the long-dash curve, then the true DUT noise temperature would have been about 9100 K, as indicated by the right-most short-dash vertical line. Obviously, a small nonlinearity in the response will result in a significant error in the measured noise temperature, and therefore the linearity must be confirmed.

It goes without saying that the system must yield repeatable measurements. This holds for the overall system, but it should also be checked for properties that are not measured every time the system is used. For example, the port reflection coefficients and asymmetries could be measured once and those values then stored and used in subsequent measurements. If that is done, it is imperative that one verify that they do not change over the time scale between measurements. Furthermore, the reliability of the measurements of reflection coefficients and S parameters should be checked.

It is also important to verify that the system is not admitting and responding to spurious signals, particularly spurious signals that vary with time. Finally, the software used to run the system and to compute the results must be verified.

The procedures for these check and verification procedures will vary for different radiometer designs and for different laboratories. Examples of the verification procedures for the coaxial and waveguide radiometers used at NIST can be found in [1, 2], respectively.

3.3 Total-Power Radiometer Uncertainties

A precision measurement is nearly worthless without an estimate of the associated uncertainty. If you truly have no estimate of the uncertainty, you could just as well have just guessed a value and saved the time and effort required to do the measurement. There are two basic approaches to estimating an uncertainty in a quantity that is computed from other measured quantities [9]. Both approaches assume knowledge of the uncertainties in the underlying measured quantities. One can either propagate uncertainties analytically, as in [10, 11], or one can perform a Monte Carlo simulation. In some cases, such as the noise-parameter measurements discussed in subsequent chapters, the determination of measured values is sufficiently complicated that the analytical approach is not practical. For the present case, where the measured noise temperature is determined by the radiometer equation (3.14), the analytical approach is relatively straightforward, and it affords more insight than the Monte Carlo approach. It is generally easier to identify the dominant contributions to the total uncertainty in the analytic approach, and thus it is easier to identify how best to improve the uncertainty.

In discussing uncertainties, we will use the following notation and conventions. Lower-case u will refer to standard (1σ) uncertainties, and upper-case U will denote expanded (2σ) uncertainties. Fractional standard uncertainties will be denoted by ξ. Nonstandard uncertainties (e.g. worst case) will be denoted by Δ. We will also follow [10] and classify uncertainties as either type-A (those determined by statistical analysis) and type-B (those determined by other means). The combined uncertainty then is the quadrature sum of the two types.

3.3.1 Type-A Uncertainties

To compute the type-A uncertainty, we need to specify the measurement procedure. For example, if one merely repeats the same measurement N times, the type-A uncertainty would be the standard deviation of the mean, that is, the sample standard deviation divided by $\sqrt{N}$,

$$u_A{}^2 = \frac{\sigma^2}{N} = \frac{1}{N(N-1)}\sum_{i=1}^{N}(\overline{T} - T_i)^2 \tag{3.15}$$

where T_i are the individual readings, and $\overline{T}$ is the mean value.

If the measurement protocol is more complicated, the statistical computation becomes more complicated. For example, for a normal calibration on the NIST coaxial radiometer [1], the DUT is measured on three different ports, and on each port 50 readings are taken. In that case, let T_{ij} be the value of a single reading, where i denotes the number of the measurement, 1 to N_M (3), and j denotes the number of the reading, 1 to N_R (50). Let T_j and σ_I refer to the average and sample standard deviation of the 50 readings in measurement I,

$$T_i = \frac{1}{N_R} \sum_{j=1}^{N_R} T_{ij}$$

$$\sigma_i^2 = \frac{1}{(N_R - 1)} \sum_{j=1}^{N_R} (T_i - T_{ij})^2 \tag{3.16}$$

and let T and σ refer to the average and standard deviation of the N_M measurements,

$$T = \frac{1}{N_M} \sum_{i=1}^{N_M} T_i$$

$$\sigma^2 = \frac{1}{(N_M - 1)} \sum_{i=1}^{N_M} (T - T_i)^2 \tag{3.17}$$

We then model T_{ij} as $T_{ij} = \tau + M_i + R_{ij}$, where $\langle M_i \rangle = \langle R_{ij} \rangle = 0$, and where τ is true value of the noise temperature, M_i is a random variable representing variations from measurement to measurement (i), and R_{ij} is a random variable the varies with the reading j and the measurement i [12]. The estimate of the true value is just the mean of all the readings,

$$\tau \approx T = \frac{1}{N_M N_R} \sum_{i=1}^{N_M} \sum_{j=1}^{N_R} T_{ij} \tag{3.18}$$

which is also equal to the mean of the N_M separate measurements. The means of the two random variables are both zero, and their variances will be denoted by $\langle M_i^2 \rangle = \nu_M$, $\langle R_{ij}^2 \rangle = \nu_R$, where the averages are over all indices. These variances can be estimated from the measured values T_{ij} by

$$\nu_R \approx \frac{1}{N_M(N_R - 1)} \sum_{i,j} (T_{ij} - T_i)^2 \tag{3.19}$$

$$\frac{1}{(N_M - 1)} \sum_{i} (T_i - T)^2 \approx \nu_M + \frac{\nu_R}{N_R} \tag{3.20}$$

The equalities are only approximate because we are dealing with a limited sample. We can solve Eqs. (3.19) and (3.20) for ν_R and ν_M, and use Eqs. (3.16) and (3.17) to write

$$\nu_R \approx \left\langle \sigma_i^2 \right\rangle, \nu_M \approx \sigma^2 - \frac{\nu_R}{N_R} \tag{3.21}$$

where the average is over the free index, and where it is understood that if a negative value results for ν_M it is taken to be 0. The type-A uncertainty in the determination of $\tau \approx T$ is then the square root of the variance in T divided by N_M,

$$u_A = \sqrt{\frac{\nu_M}{N_M} + \frac{\nu_R}{N_M N_R}} \tag{3.22}$$

with ν_R and ν_M estimated from Eq. (3.21).

This procedure can be easily generalized to more complicated cases, such as three levels of repeated measurements.

3.3.2 Type-B Uncertainties

The type-B uncertainties arise from uncertainties in the determination of the quantities appearing in the radiometer equation (3.14), and also from violations of assumptions that were made in deriving the radiometer equation. There are three such assumptions, which will be treated below.

The measured quantities appearing in the radiometer equation are: the temperature of the non-ambient standard T_S; the temperature of the ambient standard T_{amb}; the ratio of mismatch factors M_S/M_x; the ratio of the efficiencies η_{02}/η_{03}, also called the asymmetry; and the power ratios $(Y_S-1)/(Y_x-1)$. From the usual law for propagation of uncertainties [10, 11], the contribution of any one of them to the type-B uncertainty is given by

$$u_{T_x}^2 = \sum \left(\frac{\partial T_x}{\partial x_i} \right)^2 u_{x_i}^2 \tag{3.23}$$

where x_i is any of the measured quantities appearing in the radiometer equation, the subscript on the u indicates the quantity whose uncertainty it is, and we have assumed no correlations among errors in the x_i. For convenience, we shall refer to the fractional uncertainty in the quantity x_i as ξ_{x_i}.

For the non-ambient standard, the contribution to the uncertainty in T_x is given by Eq. (3.23) as

$$\frac{u_{T_x}(T_S)}{T_x} = \left| 1 - \frac{T_a}{T_x} \right| \left| \frac{T_S}{T_a - T_s} \right| \xi_{T_S} \tag{3.24}$$

For the cryogenic standards used at NIST, the fractional uncertainty in T_S, ξ_{T_S}, ranges from 0.17% to 0.77%, depending primarily on the connector type or waveguide size. For a coaxial standard in the 1–12 GHz range, $u_{T_S} \approx 0.6$ K. Thus, for a typical hot DUT, $\xi_{T_x} \approx 0.22\%$, or on the order of 20 K for a T_x of about 10 000 K.

The contribution of the ambient standard temperature is given by

$$\frac{u_{T_x}(T_a)}{T_x} = \left|\frac{T_x - T_S}{T_a - T_S}\right| \frac{T_a}{T_x} \xi_{T_a} \tag{3.25}$$

For a well-controlled ambient standard, u_{T_a} can be about 0.1 K, so that $\xi_{T_a} \approx$ 0.034%. Then for a typical hot source, ξ_{T_x} is of the order of 0.05%, which is completely negligible. Even if one is relying on the room temperature for the ambient standard, u_{T_a} could be about 1 or 2 K, in which case ξ_{T_x} would be about 0.5–1%.

The contribution to the uncertainty from the ratio of mismatch factors (denoted by M/M) is given by

$$\frac{u_{T_x}\left(\frac{M}{M}\right)}{T_x} = \left|1 - \frac{T_a}{T_x}\right| \xi_{\frac{M}{M}} \approx \left|1 - \frac{T_a}{T_x}\right| u_{\frac{M}{M}} \tag{3.26}$$

Errors in the ratio of the two mismatch factors are complicated by the correlations among errors in the various reflection coefficients in the mismatch factors. A Monte Carlo computation would handle the correlations well. In [1, 2, 8] an analytical approach was used to obtain a rather conservative, worst-case estimate of the uncertainty. Expressions were obtained for the case of all errors being perfectly correlated and for the case of them being completely uncorrelated, and the greater of the two results was used. Allowing for connector (non)repeatability and an increase in the manufacturer's specification for the uncertainty in the reflection-coefficient measurement, it was found that a typical contribution for the standard uncertainty in T_x was about 0.1–0.2%.

The uncertainty contribution from the asymmetry (denoted η/η) is given by

$$\frac{u_{T_x}\left(\frac{\eta}{\eta}\right)}{T_x} = \left|1 - \frac{T_a}{T_x}\right| \xi_{\frac{\eta}{\eta}} \approx \left|1 - \frac{T_a}{T_x}\right| u_{\frac{\eta}{\eta}} \tag{3.27}$$

There are several ways to measure the asymmetry. For the coaxial radiometer at NIST, the individual efficiencies (η_{02} and η_{03}) are measured using a reflection technique [13, 14] and the ratio is taken. On the waveguide radiometers, the asymmetry is measured by measuring two noise sources on the two measurement ports and then swapping the noise sources [2]. For the coaxial system, this results in an uncertainty in the asymmetry of between 0.0034 and 0.0042 over the 1–18 GHz range. This asymmetry uncertainty, indeed the asymmetry itself, could be eliminated by connecting all noise sources consecutively to the same measurement port. However, that would then introduce the possibility of drift between the measurements of the different sources. It would also require more effort and time.

Some uncertainty analyses for NIST radiometers [4, 8] include an explicit "connector" contribution to the uncertainty. Connector variability arises from differences between repeated connections of the same pair of connectors and from differences between different pairs of connectors, such as the difference between the DUT-VNA (vector network analyzer) connection and the DUT-radiometer connection. For the most part, connector variability is included in the uncertainty in the reflection coefficients in the mismatch and asymmetry uncertainties. It is also included (roughly) in the type-A uncertainty computed from multiple measurements with the sources disconnected between measurements. Assuming that it is included in the asymmetry and mismatch uncertainties, it is usually not necessary to introduce a separate contribution to the uncertainty due to connector variability.

For the uncertainty in the power ratios, let $Y = (Y_x - 1)/(Y_S - 1)$. Then the contribution to the uncertainty in T_x is

$$\frac{u_{T_x}(Y)}{T_x} = \left|\frac{T_S - T_a}{T_x}\right| u_Y \tag{3.28}$$

The value of u_Y will, of course, depend on the method used to measure the powers. Because Y is the ratio of a difference in measured powers, u_Y will typically be quite small. For the thermistor used in the NIST coaxial radiometers, this contribution is negligible unless $T_x \leq T_a/3$.

In addition to uncertainties in the measured quantities appearing in the radiometer equation, we must also consider possible uncertainties associated with the assumptions made in deriving that equation. One assumption was that the system was perfectly linear. The IF sections of the radiometers in [1, 2] contain circuitry for testing linearity. Provided that test is met, the contribution to the fractional uncertainty in T_x is 0.10%.

A second assumption in deriving the radiometer equation was perfect isolation. In practice, a 40 dB isolator is sufficient to ensure that the error introduced by assuming perfect isolation is less than about 0.01% [4].

There is one more potential error introduced in the derivation of the radiometer equation. All reflection coefficients and scattering parameters are measured at a single frequency, whereas the measurement of the noise power integrates over a finite bandwidth (10 MHz in the case of the NIST radiometers) centered at the nominal measurement frequency. It is possible that the reflection coefficients and scattering parameters, particularly their phases, may vary over this bandwidth. This is referred to as the "frequency offset" or "broadband mismatch" error. It can be kept negligible by judicious system design, in particular by minimizing the distance between the input port(s) and the plane at which the first significant reflection occurs [1, 8].

The total type-B uncertainty is the quadrature sum of all the aforementioned contributions,

$$u_B = \left[\begin{array}{c} u_{T_x}^2(T_S) + u_{T_x}^2(T_a) + u_{T_x}^2(Y) + u_{T_x}^2\left(\frac{M}{M}\right) + u_{T_x}^2\left(\frac{\eta}{\eta}\right) + u_{T_x}^2(isol) \\ + u_{T_x}^2(BBMM) + u_{T_x}^2(lin) \end{array} \right]^{\frac{1}{2}} \tag{3.29}$$

and the combined uncertainty is

$$u_{T_x} = \sqrt{u_A^2 + u_B^2} \tag{3.30}$$

For precision measurements, a sufficient number of repeat measurements can usually be made to render u_A small compared to u_B. For the coaxial radiometer used at NIST, the combined uncertainty is typically on the order of 0.5% for hot noise sources. Figure 3.8 shows how the fractional standard uncertainty varies as a function of the noise temperature of the DUT for the NIST coaxial radiometer at a frequency of 8 GHz and for typical reflection DUT reflection coefficient. There is a pronounced minimum when T_x is near the ambient-temperature standard noise temperature, and the fractional uncertainty rises sharply (like $1/T_x$) as T_x approaches zero.

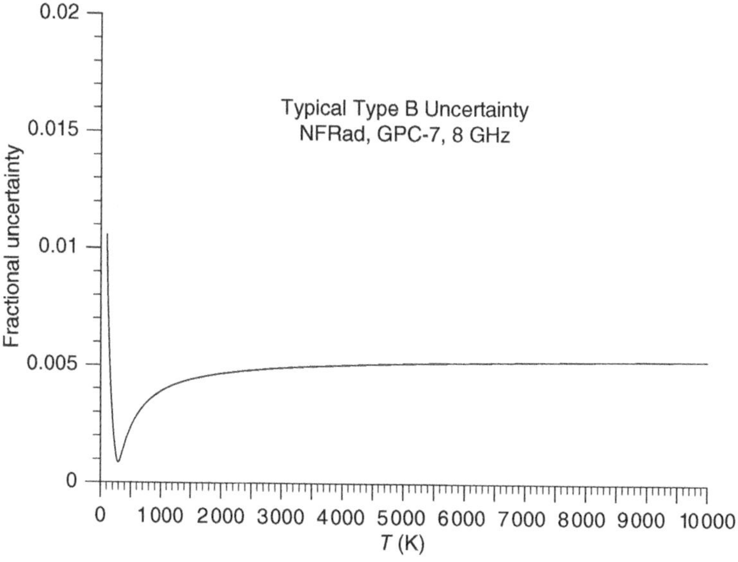

Figure 3.8 Standard fractional uncertainty as a function of DUT noise temperature at 8 GHz for the radiometer described in [1].

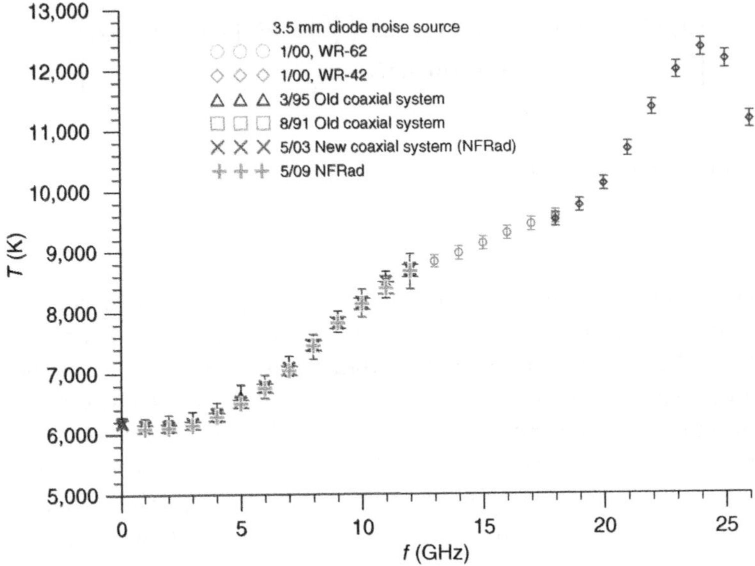

Figure 3.9 Results of noise-temperature measurements across a range of frequencies and systems.

3.3.3 Sample Results

Representative results obtained with NIST total-power radiometers are shown in Figure 3.9 for a commercial diode noise source with a PC 3.5 connector. Error bars shown are the expanded standard uncertainties, corresponding to a 95% confidence level. Besides the general size of the uncertainties, two other features are noteworthy. One is that the frequency range of the measurements extends from 30 MHz to 26 GHz and thus spans several different radiometers and primary standards. The continuity of the results as the measurement systems change is a reassuring consistency check. Also, the measurements were performed over a period of 18 years, indicating that both the measurement systems and the device are quite stable and consistent.

3.4 Other Radiometer Designs

3.4.1 Switching or Dicke Radiometer

In the early days of microwave radiometry, total-power radiometers were not practical due to the drift of the radiofrequency amplifiers that were then available. The radiometer design that was used, and that is still in use in some applications

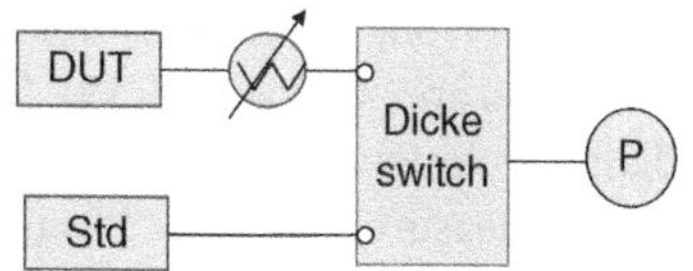

Figure 3.10 Essential components of a Dicke or switching radiometer for laboratory use.

today, was a switching radiometer, originally developed by Dicke [15]. The basic design and components of a simple realization for laboratory use are illustrated in Figure 3.10.

The DUT is compared to a known, standard noise source. The idea is to place a variable attenuator in front of the hotter of the two sources, assumed here to be the DUT, and to rapidly switch back and forth between the DUT and the standard or reference source. The "Dicke switch" switches rapidly between the two noise sources and puts out a signal proportional to the difference between them. It discriminates against any signal that is not correlated with the switching. The detection instrumentation (represented by the blob marked "P" in the figure) amplifies this difference between the powers from the two sources, and the attenuator is adjusted until the difference is zero. The DUT noise temperature is then computed from the reference temperature using Eq. (1.28).

$$\alpha_{att}T_{DUT} + (1 - \alpha_{att})T_{amb} = T_{Std} \tag{3.31}$$

where α_{att} is the available power ratio of the attenuator (equal approximately to the inverse loss), and T_{amb} is the noise temperature corresponding to the attenuator temperature (assumed to be ambient temperature).

An advantage of the switching radiometer is the relative accuracy of null detection and the absence of concerns about amplifier drift, saturation, or linearity. The rapid switching between the two sources eliminates the effect of amplifier drift, and because the difference is taken before amplification the linearity of the amplification should not be an issue. Additional sources of potential errors include the calibration of the attenuator and control and monitoring of its physical temperature. A limitation of a switching radiometer is that the DUT and the reference source must both be either hotter than ambient or colder than ambient. (Ambient here refers to the temperature of the variable attenuator.) A cryogenic standard cannot be used to measure a hot DUT, nor can a hot standard be used to measure a cold DUT. A detailed derivation of the sensitivity of a switching radiometer can be found in [16]. For laboratory use, Dicke radiometers have largely been supplanted by total-power radiometers.

3.4.2 Digital Radiometer

Digital techniques offer additional flexibility in noise measurements, particularly in enabling measurement of correlations. They have been used extensively in remote-sensing radiometry for some time [17], and also have applications in terrestrial applications [18, 19]. For purposes of noise-temperature measurements, digital techniques and instrumentation are subsumed into the unspecified output power measurement, the "P" circle in Figure 3.1 or 3.10.

3.5 Measurements through Adapters

Because it is not practical to have a radiometer for each connector type or for a radiometer to have input ports for each connector type, noise measurements must often be made through adapters. Referring to Figure 3.11, the situation is that the radiometer measures the noise temperature T_2 at plane 2, and we want to know the noise temperature of the DUT at plane 1. From Eq. (1.28),

$$T_2 = \alpha_{21} T_1 + (1 - \alpha_{21}) T_{amb} \tag{3.32}$$

and therefore

$$T_{DUT} = \frac{T_2 - (1 - \alpha_{21}) T_{amb}}{\alpha_{21}} \tag{3.33}$$

The available power ratio α_{21} is given by Eq. (1.23),

$$\alpha_{21} = \frac{|S_{21}|^2 \left(1 - |\Gamma_1|^2\right)}{|1 - \Gamma_1 S_{11}|^2 \left(1 - |\Gamma_2|^2\right)} \tag{3.34}$$

which can be determined by direct measurement of the adapter S parameters, or by other methods [14].

The flexibility afforded by using an adapter comes at a price of course, that being an increase in the uncertainty of the noise temperature measurement. From Eq. (3.33) and the usual propagation of uncertainties, we have

$$u(T_{DUT}) \approx \sqrt{\left(\frac{u(T_2)}{\alpha_{21}}\right)^2 + (T_{DUT} - T_{amb})^2 \left(\frac{u(\alpha_{21})}{\alpha_{21}}\right)^2} \tag{3.35}$$

In writing Eq. (3.35), we have neglected the contribution due to the uncertainty in T_{amb} of the adapter, because it is much smaller than the contribution from the uncertainty in α_{21}. We have therefore also neglected the effect of the correlation between errors in the measurement of T_2 and in the adapter contribution due to errors in T_{amb}.

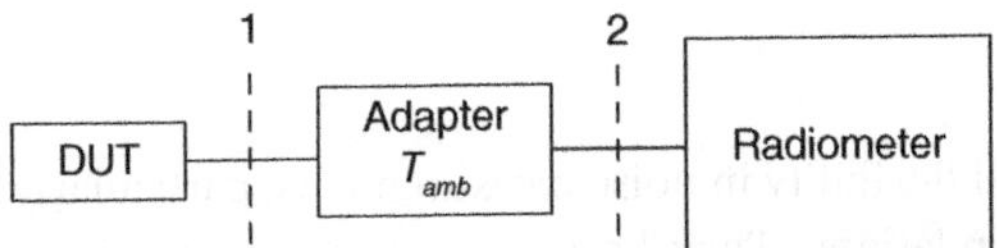

Figure 3.11 Noise-temperature measurement through an adapter.

The magnitude of the increase in the uncertainty depends on the actual values of the parameters in Eq. (3.35) of course. A good adapter will typically have a value of α_{21} between 0.90 and about 0.95 at microwave frequencies, and a good measurement of α_{21} will normally produce an uncertainty in the range 0.003–0.005. Thus, for a DUT with a noise temperature well above ambient, use of an adapter will add (in quadrature) a few tenths of a percent to the fractional uncertainty in the noise temperature measurement.

3.6 Traceability and Inter-laboratory Comparisons

In the field of thermal noise, the fundamental physical quantity is the noise temperature, which is ultimately traceable to the fundamental unit, the kelvin. Whereas the kelvin was previously defined by fixing the triple point of water, in the 2019 redefinition of the SI (Système International) the kelvin was fixed by specifying the value of the Boltzmann constant [20] to be $k_B = 1.380649 \times 10^{-23}$ J/K, where the joule (J) is defined by the fundamental units of mass, length, and time, $1\text{J} \equiv 1\text{kg} \cdot \text{m/s}^2$. Various National Measurement Institutes (NMIs) maintain realizations of the kelvin and calibrate thermometers against those realizations. Thermometers thus calibrated are directly *traceable* to the fundamentally defined kelvin. For traceable noise-temperature measurements, the measurement is traced back to a traceable temperature, such as a heated load measured with a calibrated thermometer, or the boiling point of liquid nitrogen (previously measured by calibrated thermometers), as in Chapter 2.

Since various NMIs all maintain their own realizations of the fundamental standards, including those for noise temperature, it is important that there be some means to harmonize them and to provide a mechanism whereby an NMI can check that its results are consistent with the consensus of the international community. The International Committee for Weights and Measures (CIPM) provides such a mechanism through its various consultative committees [21]. The relevant consultative committee for electromagnetic measurements is the CCEM (Consultative Committee for Electricity and Magnetism), and in particular its RF Working Group, the GTRF. They conduct international "key" comparisons and maintain an extensive database of the results, which are available on the website of the Bureau International des Poids et Mesures (BIPM) [22]. The results of such comparisons are also used to support the calibration capabilities of the NMIs.

For microwave noise, the measured quantity is noise temperature, and there are typically four to six NMIs that participate in a key comparison. Representative results from one such comparison [23], in WR-42 waveguide, are shown in Figure 3.12. Multiple devices were measured at 18, 22, 25.8, and 26.5 GHz, but results are only shown for one device. Because of different waveguide sizes only

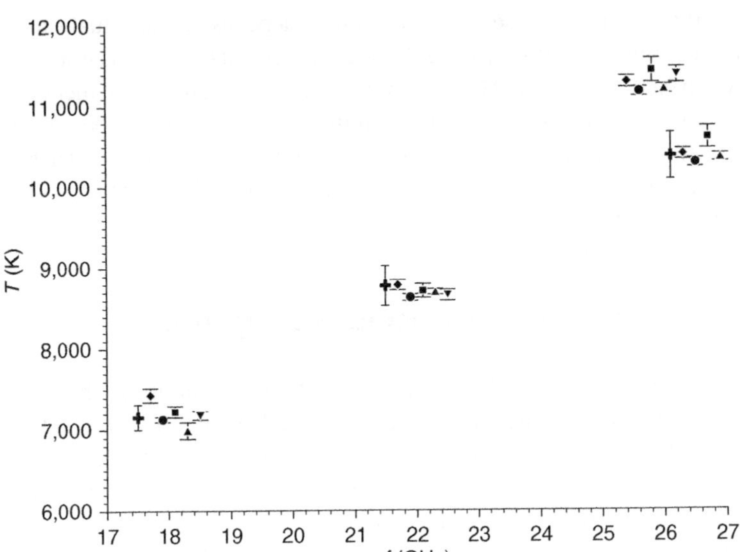

Figure 3.12 Results from a comparison of noise-temperature measurements among six NMIs. Source: Adapted from Allal [23].

five of the laboratories measured at the 25.8 and 26.5 GHz frequencies. The data points in the figure are displaced from the measurement frequencies for clarity. For full details of the comparison and the analysis of the data, see [23].

References

1 C.A. Grosvenor, J. Randa, and R.L. Billinger, "Design and testing of NFRad—A new noise measurement system," National Institute of Standards and Technology (U.S.) Technical Note 1518 (March 2000). https://tsapps.nist.gov/publication/get_pdf.cfm?pub_id=6264 (Accessed 17 March, 2022).

2 J. Randa and L.A. Terrell, "Noise-temperature measurement system for the WR-28 band," National Institute of Standards and Technology (U.S.) Technical Note 1395 (August 1997). https://tsapps.nist.gov/publication/get_pdf.cfm?pub_id=20255 (Accessed 17 March, 2022).

3 C.A. Grosvenor and R.L. Billinger, "The NIST 30/60 MHz tuned radiometer for noise temperature measurements," NIST Technical Note TN1525 (May 2002).

4 W.C. Daywitt, "Radiometer equation and analysis of systematic errors for the NIST automated radiometers," National Institute of Standards and Technology (U.S.) Technical Note 1327 (March 1989).

5 G.L. Williams, "Source mismatch effects in coaxial noise source calibration," *Measurement Science and Technology*, **2**, no. 8, pp. 751–756 (1991).

6 M.W. Sinclair, "Untuned systems for the calibration of electrical noise sources," *IEE Colloquium on 'What's New in Microwave Measurements'* (Digest No: 1990/174), pp. 7/1–7/5 (1990).

7 D.F. Wait, "Radiometer equation for noise comparison radiometers," *IEEE Transactions on Instrumentation and Measurement*, **44**, no. 2, pp. 336–339 (April 1995).

8 J. Randa, "Uncertainties in NIST noise-temperature measurements," National Institute of Standards and Technology (U.S.) Technical Note 1502 (March 1998). doi: https://doi.org/10.6028/NIST.TN.1502.

9 ISO/IEC Guide 98-3:2008, Uncertainty of measurement—Part 3: Guide to the uncertainty in measurement (GUM:1995), (2008).

10 ISO *Guide to the Expression of Uncertainty in Measurement*, International Organization for Standardization, Geneva, Switzerland, 1993.

11 B.N. Taylor and C.E. Kuyatt, "Guidelines for evaluating and expressing the uncertainty of NIST measurement results," National Institute of Standards and Technology (U.S.) Technical Note 1297, 1994 edition (September1994).

12 F.A. Graybill, *Theory and Application of the Linear Model*, Duxberry Press, Belmont, CA, 1976.

13 W.C. Daywitt, "Determining adaptor efficiency by envelope averaging swept frequency reflection data," *IEEE Transactions on Microwave Theory and Techniques*, **38**, no. 11, pp. 1748–1752 (November 1990).

14 J. Randa, W. Wiatr, and R.L. Billinger, "Comparison of adapter characterization methods," *IEEE Transactions on Microwave Theory and Techniques*, **47**, no. 12, pp. 2613–2620 (December 1999). doi: 10.1109/22.750234.

15 R.H. Dicke, "The measurement of thermal radiation at microwave frequencies," *Review of Scientific Instruments*, **17**, no. 7, pp. 268–275 (1946). doi: 10 .1063/1.1770483. PMID 20991753.

16 D.F. Wait, "The sensitivity of the Dicke radiometer," *NIST Journal of Research*, **71C**, no. 2, pp. 127–152 (April 1967). doi: https://doi.org/10.6028/jres.071C.015.

17 C. Ruf and S. Gross, "Digital radiometers for earth science," in *2010 IEEE MTT-S International Microwave Symposium*, Anaheim, CA, 2010, pp. 828–831. doi:10.1109/MWSYM.2010.5518137.

18 I. Rolfes, T. Musch, and B. Schiek, "A highly linear digital detector for noise parameter measurements at microwave frequencies," in *2001 31st European Microwave Conference*, London, 2001, pp. 1–4. doi: 10.1109/EUMA.2001.338940.

19 X. Lu, D. Kuester, and D. Gu, "Digital radiometer for traceable spectrum sensing," in *2020 Conference on Precision Electromagnetic Measurements (CPEM)*, Denver, CO, 2020, pp. 1–2. doi: https://doi.org/10.1109/CPEM49742.2020 .9191911.

20 D.B. Newell , F. Cabiati, J. Fischer, K. Fujii, S.G. Karshenboim, H.S. Margolis, E. de Mirandés, P.J. Mohr, F. Nez, K. Pachucki, T.J. Quinn, B.N. Taylor, M. Wang, B.M. Wood, and Z. Zhang, "The CODATA 2017 values of h, e, k, and N_A for the revision of the SI," *Metrologia*, **55**, no. 1 pp. L13 (April 2018). doi: https://doi.org/10.1088/1681-7575/aa950a.

21 https://www.bipm.org/en/committees/ci/cipm (Accessed 10 March, 2022).

22 https://www.bipm.org/kcdb (Accessed 10 March, 2022).

23 D. Allal, "CCEM RF Key Comparison CCEM.RF-K22.W, Noise in waveguide between 18 GHz and 26.5 GHz, Final Report," (April, 2016). Online: https://www.bipm.org/documents/20126/45255714/CCEM.RF-K22.W_Final_Report.pdf/9a946c97-0ffd-7e23-ac5a-17ab54f45672?version=1.1&t=1603457339532 (Accessed 10 March, 2022).

4

Amplifier Noise

4.1 Noise Figure, Effective Input Noise Temperature

Characterization of the noise added to a signal by an amplifier is one of the most important applications of microwave noise. The quantity that is usually used to measure the noise added by an amplifier is the *noise factor*, F, which is defined as [1]: (at a given frequency) the ratio of total output noise power per unit bandwidth to the portion of the output noise power which is due to the input noise, evaluated for the case where the input noise power is $k_B T_0$, where $T_0 = 290$ K. If we refer to Figure 4.1, we see that the noise factor is a measure of the degradation of the input signal-to-noise ratio due to the amplifier [2],

$$\frac{(S/N)_{in}}{(S/N)_{out}} = \frac{\dfrac{S_{in}}{290\,K}}{\dfrac{(GS_{in})}{(G \times 290\,K + N_{amp})}} = \frac{G \times 290\,K + N_{amp}}{G \times 290\,K} = F \tag{4.1}$$

where G is the amplifier gain, and N_{amp} is the noise (power spectral density) added by the amplifier. Because we are dealing with noise temperatures, which are defined as available powers, the gain G is the *available* gain. A distinction is sometimes made between the noise factor and the *noise figure*, which is the noise factor expressed in dB. We see little point in making this distinction and will use the two interchangeably.

Since it is common to work with noise powers expressed as noise temperatures, it is convenient to define an *effective input noise temperature*, T_e, associated with a noise factor. It is defined as the input noise temperature that would result in the observed output noise power spectral density due to the amplifier,

$$N_{amp} \equiv G k_B T_e \tag{4.2}$$

Precision Measurement of Microwave Thermal Noise, First Edition. James Randa.

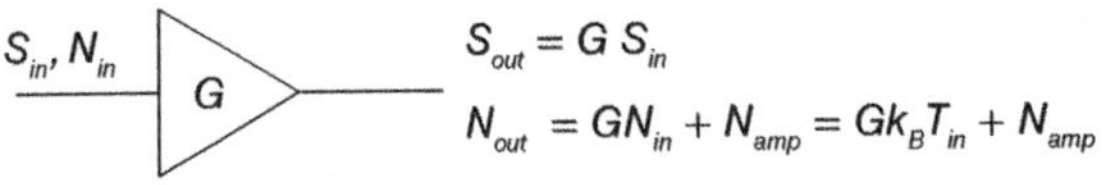

S_{in}, N_{in} → G → $S_{out} = G\,S_{in}$

$N_{out} = GN_{in} + N_{amp} = Gk_BT_{in} + N_{amp}$

Figure 4.1 Input and output signal and noise of an amplifier.

Thus, referring again to Figure 4.1, $N_{out} = Gk_B(T_{in} + T_e)$, and consequently

$$F = \frac{G(T_0 + T_e)}{GT_0}, \quad F(dB) = 10log_{10}\left(\frac{T_0 + T_e}{T_0}\right) \tag{4.3}$$

It is important to note that the quantities F, G, and T_e all depend on the reflection coefficient of the source that is connected to the input of the amplifier. Thus, if one refers to *the* noise figure, it is necessary to specify the source reflection coefficient that is assumed (matched, optimal, etc.).

4.2 Noise-Temperature Definition Revisited

This is an appropriate point to return to the definition of noise temperature which was discussed in Section 1.3. There we considered two possible ways to define the noise temperature – either as the physical temperature that would result in the observed available power, or as the available power spectral density divided by k_B. We chose to define the noise temperature as the available power spectral density divided by k_B, which allows us to write equations like

$$N_{out} = Gk_B(T_{in} + T_e) \tag{4.4}$$

where G is the available gain. If instead we had chosen the "equivalent physical temperature" definition, then $N_{in} = hf/\left(e^{hf/k_BT_{in}} - 1\right)$ and similarly for the other noise temperatures. With that definition, Eq. (4.4) would instead become

$$\frac{hf}{e^{hf/k_BT_{out}} - 1} = G\left(\frac{hf}{e^{hf/k_BT_{in}} - 1} + \frac{hf}{e^{hf/k_BT_e} - 1}\right)$$

Solving for T_{out} would then yield

$$T_{out} = \frac{hf}{k_B}\left\{ln\left[1 + \frac{1}{G}\left(\frac{1}{\left(e^{hf/k_BT_{in}} - 1\right)} + \frac{1}{\left(e^{hf/k_BT_e} - 1\right)}\right)^{-1}\right]\right\}^{-1}$$

rather than

$$T_{out} = G(T_{in} + T_e) \tag{4.5}$$

as it is with the available-power definition that we adopted. As was discussed in Chapter 1, for most physical temperatures and microwave frequencies, the difference does not matter. For precise measurements at low temperature and/or high frequency, however, there is a (small) difference.

4.3 Noise Figure Measurement, Simple Case

In discussing the measurement of the noise figure of an amplifier, we first consider a simplified case, in order to introduce some notation and concepts, and also to establish a baseline that will be useful for intuition in the more complicated general case. We assume lossless lines and single-mode transmission in lines, those assumptions are standard and usually quite justifiable. We also assume perfect matching (all reflection coefficients and diagonal S-matrix elements = 0). Referring to Figure 4.2, two noise sources ("hot" and "cold") with known noise temperatures (T_h and T_c) are connected sequentially to the amplifier input, and the output noise power density is measured for each. This results in two equations,

$$N_{out,h} = Gk_B(T_h + T_e), \quad N_{out,c} = Gk_B(T_c + T_e) \tag{4.6}$$

which can be combined and solved to yield

$$T_e = \frac{T_h - YT_c}{Y - 1}, \qquad G = \frac{N_{out,h} - N_{out,c}}{k_B(T_h - T_c)} \tag{4.7}$$

where $Y = N_{out,h}/N_{out,c}$. The noise figure is then given by

$$F = 1 + \frac{T_e}{T_0} = 1 + \frac{T_h - YT_c}{(Y - 1)T_0} \tag{4.8}$$

For obvious reasons, this is known as the Y-factor method of measuring the noise figure. Although we assumed reflectionless input sources, the Y-factor method can be used as long as the reflection coefficients of the two input sources are *equal*. The two input noise temperatures do not need to be hot or cold in any absolute sense; they just need to be different, so that one is greater than the other, $T_h > T_c$. A common practice is to use a noise diode source in its "on" state as the hot source and the same noise diode source in its "off" state as the cold source. One can then switch between the hot and cold sources simply by turning on and off the power supply to the diode. When the power is off, the inactive noise diode source is simply a passive termination whose noise temperature is given by Eq. (1.6), with T_{phys} equal to the ambient temperature (assuming the noise source is in equilibrium with its surroundings). This practice relies on the diode noise

Figure 4.2 Measurement of noise factor or T_e.

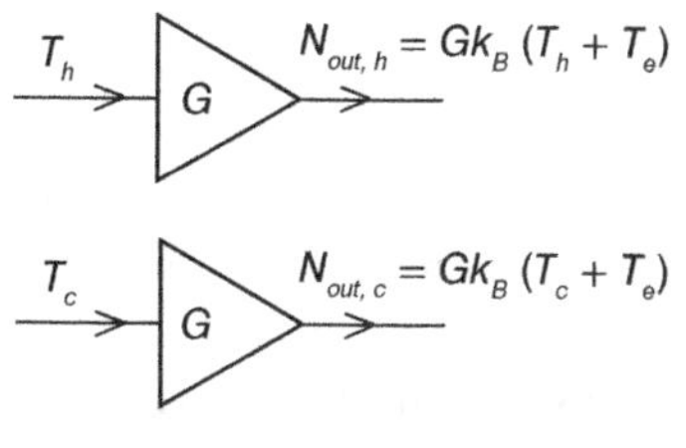

T_h → G → $N_{out,h} = Gk_B\,(T_h + T_e)$

T_c → G → $N_{out,c} = Gk_B\,(T_c + T_e)$

source being well matched in both its on and off states (It also requires that the amplifier continue to behave linearly for both input terminations.)

Equation (4.7) indicates that this method of measuring the noise figure or effective input noise temperature also results in a measurement of the available gain, due to their entanglement in the definition of T_e, $T_{out} = G(T_{in} + T_e)$, where G sets the scale of the relationship between the known input noise and the output noise temperature (or power). If G is known from some other measurement, such as vector network analyzer (VNA) measurements, then a single noise measurement with a known input noise temperature would suffice to determine T_e or F, as can be seen from Eq. (4.6). Not surprisingly, something similar to this feature occurs in the measurement of noise parameters as well.

If one is only interested in crude computations, there are approximations that are sometimes used to express the noise figure in terms of the power ratio Y and the *ENR* (Eq. 1.7) of the hot source. We first write the noise factor in terms of the *ENR*,

$$F = 1 + \frac{T_e}{T_0} = 1 + \frac{T_h - YT_c}{(Y-1)T_0} = \frac{ENR_h}{(Y-1)} + \left(\frac{Y}{Y-1}\right)\left(\frac{T_0 - T_c}{T_0}\right) \tag{4.9}$$

If the cold source has a noise temperature near T_0, as is the common case in which it is at room temperature, the second term on the right in Eq. (4.9) is small, and

$$F \approx \frac{ENR_h}{(Y-1)}, \quad F(dB) \approx ENR_h(dB) - (Y-1)(dB) \tag{4.10}$$

where $(Y-1)(dB)$ is the quantity $(Y-1)$ expressed as dB. It should be noted that $T_0 = 290\,\text{K}$ is about 63 °F, which is cooler than a typical laboratory ambient temperature. Approximations such as Eq. (4.10) can be useful in conversation or for rough estimates and mental computations. For any "real" computation, however, one should use the full, correct expression(s). It only takes a few seconds of extra typing, and it can make a difference in the answer.

4.4 Definition of Noise Parameters

4.4.1 Circuit Treatment of Noisy Amplifier

The discussion in Section 4.3 was restricted to the idealized matched case. As is to be expected, things are more complicated in the real world. Early treatments of noisy amplifiers were based on Thevinin-equivalent circuit representations, and those treatments still underpin noise-parameter work today. There are several different, but equivalent, Thevinin-equivalent circuits for a noisy amplifier (e.g. see [3–6]). The one shown in Figure 4.3 consists of a noiseless amplifier with noise

Figure 4.3 Equivalent circuit for an amplifier with noise.

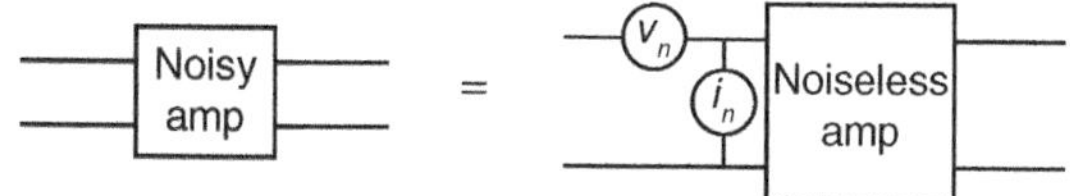

voltage (v_n) and current (i_n) sources on the input. The mean-square averages of v_n and i_n, along with their (complex) cross correlation $\langle v_n i_n^* \rangle$, are sufficient to characterize the internal noise sources of the amplifier, but that is not the most convenient parameterization. In the Rayleigh-Jeans approximation, the mean square open-circuit noise voltage across a resistance R, at temperature T, is given by $\langle |v_n|^2 \rangle = 4k_B TR\Delta f$. We then define *equivalent noise resistance* R_n to be

$$R_n \equiv \frac{\langle |v_n|^2 \rangle}{4k_B T_0 \Delta f} \tag{4.11}$$

which is the value of the resistance at temperature T_0 that would give rise to the $\langle |v_n|^2 \rangle$ of the amplifier. This provides a convenient means for comparing the magnitude of the amplifier noise to noise generated by resistances at the reference temperature. In a similar manner, the *equivalent noise conductance* G_n is defined as

$$G_n \equiv \frac{\langle |i_n|^2 \rangle}{4k_B T_0 \Delta f} \tag{4.12}$$

If one then considers the noise figure, it is possible to put it into the form [7]

$$F = F_{min} + \frac{R_n}{G_S} |Y_{opt} - Y_S|^2 \tag{4.13}$$

where Y_S is the admittance of the source, G_S is its conductance, F_{min} is the minimum possible value for F as a function of Y_S and G_S, and Y_{opt} is the source admittance for which this minimum occurs. In this representation the four *noise parameters* are F_{min}, R_n, and the complex Y_{opt}. They are the four parameters that completely characterize the noise behavior of the amplifier. If desired, the quantities $\langle |v_n|^2 \rangle$, $\langle |i_n|^2 \rangle$, and $\langle v_n i_n^* \rangle$ can be computed from the noise parameters [7], but it is often sufficient to work with the noise parameters themselves, and they are the quantities that are most often measured. Taken together, Eqs. (4.3) and (4.13) imply a similar parameterization for the effective input noise temperature T_e,

$$T_e = T_{e,min} + \frac{R_n T_0}{G_S} |Y_{opt} - Y_S|^2 \tag{4.14}$$

where $T_{e,\,min} = (F_{min} - 1)T_0$. We can also write T_e in terms of the source reflection coefficient Γ_S, rather than the source admittance,

$$T_e(\Gamma_S) = T_{e,min} + \frac{4R_n T_0}{Z_0} \frac{|\Gamma_{opt} - \Gamma_S|^2}{|1 + \Gamma_{opt}|^2 \left(1 - |\Gamma_S|^2\right)} \tag{4.15}$$

where Z_0 is a reference impedance, usually taken to be 50 Ω. Eqs. (4.13)–(4.15) will be referred to as the *IEEE representation* of the noise parameters [7].

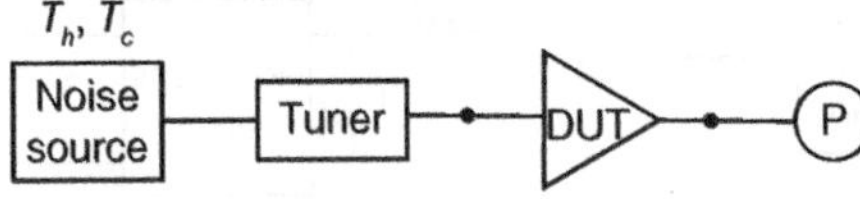

Figure 4.4 Measurement setup for measurement of noise parameters.

Equation (4.13) shows explicitly that the noise figure depends on the input impedance. It also suggests a direct way to measure the noise parameters, which was in fact the initial method used [8], before computers transformed measurement and calculation technology. The measurement setup that was employed in the early measurements is shown in Figure 4.4. The tuner allowed measurement of the output noise power for two different input noise temperatures at the same source admittance, so that the noise figure at that value of source admittance could be determined from the Y-factor method. The input noise sources were often taken to be the on and off states of a noise diode source, as discussed above. The tuner was then used to vary the source admittance (T_h and T_c at same admittance), measuring the noise figure for different source admittances until the minimum value of the noise figure was found. That determined the value of F_{min}, and the admittance at which this minimum occurred was Y_{opt}. To determine R_n, the source impedance was varied, and the paraboloid around Y_{opt} was fit.

4.4.2 Wave Representation of Noise Parameters

For microwave radiometry, a wave or noise-matrix representation of the noise parameters often provides more flexibility. The usual S-matrix equation is modified to include the contribution of outgoing noise waves due to the amplifier itself, and a linear two-port is described by

$$\begin{pmatrix} b_1 \\ b_2 \end{pmatrix} = \begin{pmatrix} S_{11} & S_{12} \\ S_{21} & S_{22} \end{pmatrix} \begin{pmatrix} a_1 \\ a_2 \end{pmatrix} + \begin{pmatrix} c_1 \\ c_2 \end{pmatrix} \tag{4.16}$$

where the amplitudes c_1 and c_2 are the result of intrinsic noise sources within the two-port device [9–14]. If a noise source is connected to the input of an amplifier, as in Figure 4.5, the output noise power density is given by

$$T_2 = GT_S + \frac{|S_{21}|^2}{(1 - |\Gamma_2|^2)}(N_1 + N_2 + N_{12}) \tag{4.17}$$

Figure 4.5 Source and amplifier in the wave representation.

where G is the available gain, given by Eq. (1.23), and Γ_2 is given by Eq. (1.24). The terms N_1, N_2, and N_{12} are due to the amplifier noise and are given by

$$N_1 = \left| \frac{\Gamma_S}{1 - \Gamma_S S_{11}} \right|^2 X_1,\ N_2 = X_2,\ N_{12} = 2Re\left[\frac{\Gamma_S}{(1 - \Gamma_S S_{11})} X_{12} \right] \tag{4.18}$$

where we have introduced the notation

$$k_B X_1 \equiv \left\langle |c_1|^2 \right\rangle, \quad k_B X_2 \equiv \left\langle |c_2/S_{21}|^2 \right\rangle, \quad k_B X_{12} \equiv \langle c_1 (c_2/S_{21})^* \rangle \tag{4.19}$$

In deriving Eq. (4.17) we have used the fact that $\left\langle c_S c_1^* \right\rangle = \left\langle c_S c_2^* \right\rangle = 0$, where c_S is the c_G in Eqs. (1.16–1.18), since the noise wave from the termination is not correlated with those from the amplifier. These X parameters are distinct from, and should not be confused with, other sets of X parameters, such as those introduced in the analysis of nonlinear circuit behavior.

If we consider that c_1 and c_2 are due to the intrinsic amplifier noise emerging respectively from the amplifier input and output ports, we see that in Eq. (4.18), N_1 arises from the noise wave that emerges from the input of the amplifier, gets reflected from the source back into the amplifier, and is amplified and appears at the output. N_2 is due to the noise wave that emerges directly from the output of the amplifier. N_{12} is due to the interference and correlation of c_1 (after reflection and amplification) with c_2 at the output.

As in the circuit treatment above, we see that the output noise power or noise temperature depends on the reflection coefficient of the input termination. Consequently, the effective input noise temperature of the amplifier must also depend on the input termination, since $T_2 = G(T_{in} + T_e)$. If we force Eq. (4.17) into the form $T_2 = G(T_{in} + T_e)$, we can identify T_e as

$$T_e = \frac{1}{\left(1 - |\Gamma_S|^2\right)} \left\{ |\Gamma_S|^2 X_1 + |1 - \Gamma_S S_{11}|^2 X_2 + 2Re\left[(1 - \Gamma_S S_{11})^* \Gamma_S X_{12}\right] \right\} \tag{4.20}$$

The X parameters are a representation of the noise parameters in the wave representation. From Eq. (4.19) we see that they have dimensions of temperature. In this representation, the four noise parameters are X_1, X_2, and the real and imaginary parts of X_{12}.

The circuit (or IEEE) representation and the wave representation of the noise parameters are equivalent, and if the reflection coefficients and S parameters are known, one set can be written in terms of the other. If the IEEE noise parameters are known, the wave representation parameters can be computed from [15]:

$$X_1 = T_{e,min}\left(|S_{11}|^2 - 1\right) + \frac{t|1 - S_{11}\Gamma_{opt}|^2}{|1 + \Gamma_{opt}|^2}, \quad X_2 = T_{e,min} + \frac{t|\Gamma_{opt}|^2}{|1 + \Gamma_{opt}|^2},$$

$$X_{12} = S_{11} T_{e,min} - \frac{t\Gamma_{opt}^*(1 - S_{11}\Gamma_{opt})}{|1 + \Gamma_{opt}|^2} \tag{4.21}$$

where we have introduced the symbol $t \equiv (4R_n T_0/Z_0)$. Conversely, if the wave parameters are known, the IEEE parameters can be computed from

$$T_{e,min} = \frac{X_2 - |\Gamma_{opt}|^2 \left[X_1 + |S_{11}|^2 X_2 - 2Re\left(S_{11}^* X_{12}\right)\right]}{\left(1 + |\Gamma_{opt}|^2\right)}$$

$$t = X_1 + |1 + S_{11}|^2 X_2 - 2Re\left[(1 + S_{11})^* X_{12}\right]$$

$$\Gamma_{opt} = \frac{\eta}{2}\left(1 - \sqrt{1 - \frac{4}{|\eta|^2}}\right) \tag{4.22}$$

where

$$\eta = \frac{X_2\left(1 + |S_{11}|^2\right) + X_1 - 2Re\left(S_{11}^* X_{12}\right)}{(X_2 S_{11} - X_{12})} \tag{4.23}$$

A few features of this representation of the noise parameters and the relationships between the two sets of noise parameters are worth noting. From Eq. (4.19) we see that the effective input noise temperature for the matched case ($\Gamma_S = 0$), sometimes called the "50 Ω noise temperature," is given by $T_{e,0} = X_2$. Thus, X_2 could be measured directly by the Y-factor method described in Section 4.2. X_1 is the intrinsic noise emerging from the input port (1). It is closely related to the T_{rev} of Wait and Engen [13],

$$T_{rev} = \frac{X_1}{\left(1 - |S_{11}|^2\right)} \tag{4.24}$$

And it can be measured directly from the configuration shown in Figure 4.6. (This is discussed further in Section 4.5.4.) In principle, X_{12} can also be measured directly, by combining the input and output ports of the amplifier with different phases. Therefore, all four of the noise parameters in the wave representation can be measured directly [15], but such a direct measurement requires the ability to measure $\langle c_1 (c_2/S_{21})^* \rangle$, the correlation of the intrinsic noise waves emerging from the two ports. As the phase noise of new instrumentation decreases, the measurement of such correlations becomes more accurate, and the direct measurement of the wave-representation noise parameters could become more common.

A noteworthy feature of the relationships in Eqs. (4.21)–(4.23) is that they enable us to derive physical bounds on various combinations of the noise parameters. For example, since X_1 must be greater than or equal to zero, we have

$$T_{e,min} \leq \frac{t|1 - S_{11}\Gamma_{opt}|^2}{\left(1 - |S_{11}|^2\right)|1 + \Gamma_{opt}|^2} \tag{4.25}$$

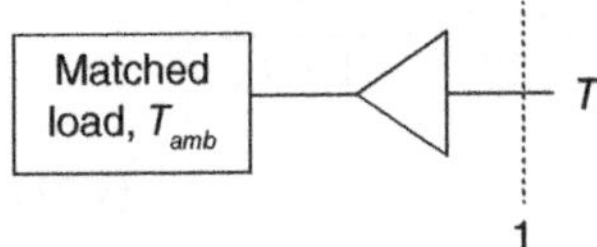

Figure 4.6 Configuration for measurement of reverse noise.

Similarly, applying the Schwarz inequality to Eq. (4.20) yields

$$X_1 + X_2 \geq 2|X_{12}| \tag{4.26}$$

It is also possible to show that

$$|\eta| \geq 2 \tag{4.27}$$

which constrains the noise parameters through Eq. (4.23). We will revisit these bounds when we discuss verification and checks below.

4.5 Measurement of Noise Parameters

4.5.1 General Measurement Setup

There are many different methods for measuring amplifier (or other linear two-port) noise parameters. Some of the earlier methods are found in [11–13, 15–25]. Most of them are based on the IEEE parameterization. The basic idea of (almost) all methods is to start with an equation for a measurable output noise quantity (such as the noise power or temperature, the noise figure, or the effective input noise temperature) in terms of the noise parameters and various input quantities, such as the input noise power and reflection coefficient and the DUT S-parameters. Equations (4.9), (4.12), and (4.14) are examples of such equations. The DUT is then presented with a variety of different input reflection coefficients and noise temperatures, and the output is measured for each. This yields a series of equations involving the noise parameters and known (previously measured) quantities. Usually, more input terminations are used than the minimum that would be needed, and thus the system of equations is overdetermined. It is solved by a least-squares fit for the noise parameters. Considerable work has been done on choosing an optimal set of input reflection coefficients, but generally it is sufficient to have a set that exhibits good coverage of the complex plane. This will be discussed further in Section 4.5.5. In addition to the four noise parameters, a gain parameter can be determined by the fit, or it can be taken from S-parameter measurements.

A representation of a generic measurement setup for the methods of this section is shown in Figure 4.7. The key components of the system comprise the set of input terminations, the VNA, the output measurement system, and of course the DUT. We will discuss each of these in turn, including the DUT, which for the purposes of this section is assumed to be a packaged amplifier. We will not dwell on details of instrumentation, but rather on the functions of the various components.

The input terminations comprise devices or states (denoted by i) with known noise temperature (T_i) and reflection coefficient (Γ_i), which are connected

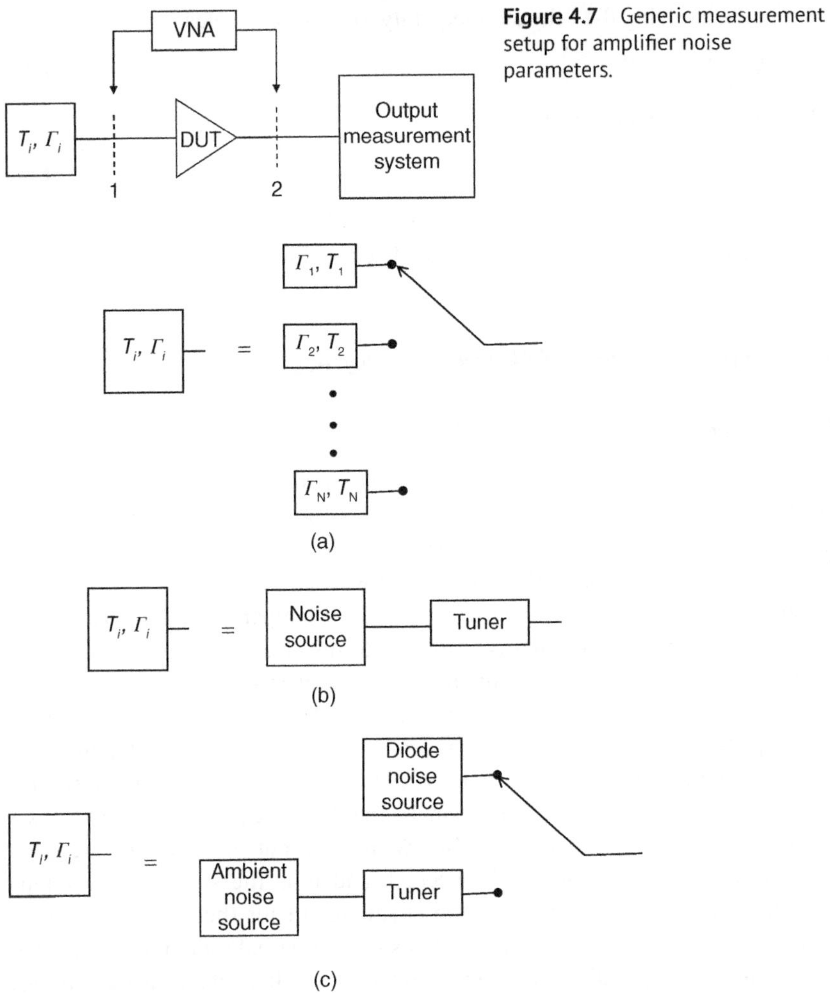

Figure 4.7 Generic measurement setup for amplifier noise parameters.

Figure 4.8 (a) Discrete terminations for input states. (b) Use of a tuner to achieve different input states. (c) Alternative configuration in which tuner is bypassed for the hot input termination.

sequentially to the input of the DUT. As indicated in Figure 4.8a,b, the different states can be achieved by switching among numerous discrete terminations, or a tuner can be used with an input noise source. If a tuner is used, the input noise source can be a diode, whose noise temperature can be varied by turning the diode on and off. If a tuner is used, it is necessary to account for its effect in computing the noise temperature at the input of the amplifier [26] (An alternate configuration that employs a tuner is shown in Figure 4.8c. This configuration can be used if one fits to noise temperature, rather than noise figure, as discussed in Sections 4.5.2 and 4.5.3.) If no tuner is used, the set of discrete states can be achieved by using a commercial electronic calibration unit or by simply assembling a battery of terminations with appropriate switching [27]. The crucial point is that the noise temperature and reflection coefficient of each state are known, and an obvious prerequisite is that both noise temperature and reflection coefficient be stable and repeatable. In order to control the noise temperature of passive devices, it is important that their physical temperature be stable over the course of the measurement. The same is true for diode noise sources, since their noise temperature varies with their physical temperature.

In Figure 4.7, the VNA is used to measure the various reflection coefficients and scattering parameters that are needed for the measurements. These include the reflection coefficients $\Gamma_{1,i}$ of the input terminations or states at the input to the DUT (plane 1), the S-parameters of the DUT, the input reflection coefficient of the output measurement system, and the reflection coefficients $\Gamma_{2,i}$ looking into the output of the DUT for each input termination. The $\Gamma_{2,i}$ can either be measured directly, or they can be calculated from the measured $\Gamma_{1,i}$ and the S-parameters. However, more accurate results can usually be achieved by direct measurement of the $\Gamma_{2,i}$.

The output measurement system usually consists of some variation on the theme of an RF receiver, intermediate-frequency (IF) amplification, and some sort of power meter, P, as indicated in Figure 4.9a. This is the same basic system that was treated in the discussion of the total-power radiometer in Section 3.1. Here we present an equivalent, but slightly different, analysis. The system is designed to have a linear response, so that the output power p_{out}^{del} is given by a constant term

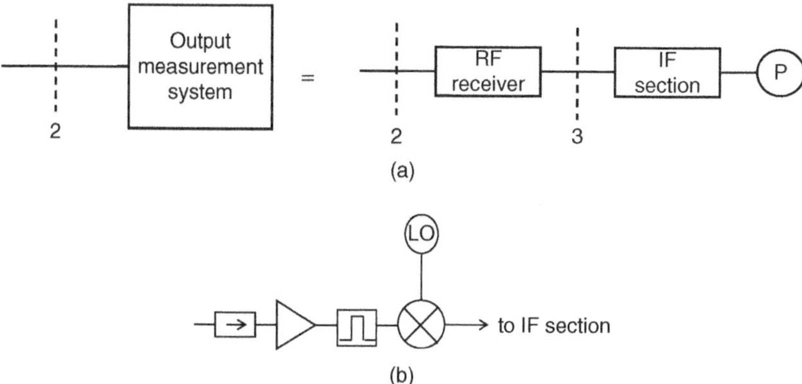

Figure 4.9 (a) General form for the output measurement system. (b) RF receiver for output measurement system, including optional isolator.

due to the receiver noise plus a term proportional to the input power p_{in}^{del},

$$p_{out}^{del} = a + bp_{in}^{del} \tag{4.28}$$

The calibration constants a and b can be related to the receiver's noise figure and gain, respectively, and like the noise figure and gain, they are constant with respect to the power, but they depend on the various reflection coefficients. To derive the relationship, consider a reference plane 3 at the output of the receiver. The power delivered by the receiver to the IF section and the power meter can be written as the available power at plane 3 times the mismatch factor there, and the available power at the receiver's output is related to its available gain G_{rec}^{av} and effective input noise temperature T_{rec}, leading to

$$p_3^{del} = M_3 p_3^{av} = M_3 G_{rec}^{av} k_B (T_{rec} + T_2) \tag{4.29}$$

Comparison of Eqs. (4.28) and (4.29) then indicates that $a = M_3 G_{rec}^{av} k_B T_{rec}$, and $b = (M_3/M_2) G_{rec}^{av}$. Note that we can write things in terms of either the spectral power delivered to the receiver (p_2^{del}) or the noise temperature at plane 2 ($p_2^{av} = k_B T_2$); the difference lies in the inclusion (or exclusion) of the appropriate mismatch factor M_2.

Like the noise figure and gain, a and b depend on the reflection coefficient of the source connected to the receiver's input, $\Gamma_{2,i}$. As in Chapter 3, there are two ways of handling this dependence in the calibration of the output measurement system. One way is to include an isolator at the input to the receiver, thus rendering the receiver's response, as well as the values of a and b, independent of $\Gamma_{2,i}$. The values of a and b for any value of $\Gamma_{2,i}$ can then be determined by measuring the output for just two different values of p_{in}. The amount of isolation required depends on the desired uncertainty, but 40 dB is usually sufficient. Of course, the simplicity gained by using an isolator comes at a price. Isolators are effective only over a limited bandwidth, thus restricting the bandwidth of the output measurement system (An isolator also introduces additional loss, which reduces the resolution of the measurement system, but this is typically not a significant problem.) The restriction imposed by the limited bandwidth of an isolator can be overcome by using several isolators, covering different frequency bands, in parallel and switching among them to choose the one appropriate to the frequency of interest [28].

The other way to deal with the dependence on $\Gamma_{2,i}$ is to measure the noise parameters of the receiver, so that the receiver gain and noise temperature can be calculated for any value of the input reflection coefficient $\Gamma_{2,i}$ [17, 29]. This is done in essentially the same manner as will be described in Sections 4.5.2–4.5.7, except that the receiver itself serves as both the DUT and the output measurement system.

The general form of a receiver used in the output measurement system is shown in Figure 4.9b. The isolator may be present or not, as just discussed. The RF amplifier is usually a low-noise amplifier (LNA), although for measurements on amplifiers the input to the receiver is the output of an amplifier and thus does not necessarily impose very stringent demands on the resolution. The input to the receiver is broadband, and so a filter is used to restrict the range of frequencies input to the mixer. If a band-pass filter with pass band centered at f_0 and bandwidth Δf is used, the mixer output will comprise two sidebands of width Δf, centered at $f_{LO} \pm f_0$, where f_{LO} is the local oscillator (LO) frequency. If a low-pass filter is used, the output will again be centered at the LO frequency, but will extend continuously from $f_{LO} - \Delta f$ to $f_{LO} + \Delta f$. A DC block is often inserted into the IF chain to remove any stray DC component. The noise measurements represent an average over the bandwidth of the filter.

4.5.2 Fit to Noise-Figure Parameterization

Many different strategies have been employed to measure the noise parameters of packaged amplifiers. As discussed in Section 4.4.1, early measurements employed an experimental search for the minimum noise figure, thus determining F_{min} and Γ_{opt}, and then varied the input impedance around this minimum to determine R_n. As computing power grew and became more available, the methods evolved. The most popular current methods are based on measurements of the form represented in Figure 4.7. A series of input terminations with known noise temperature and reflection coefficient are connected to the amplifier input; and the output noise power, noise temperature, or noise figure are measured for each. The expression for the measured quantity as a function of the noise parameters is then fitted to the set of measurement results. The different measurement strategies are distinguished by features of the set of input states, how the different input states are achieved, and the quantity that is measured (and fitted).

Initial fitting work on noise-parameter measurements was based on measurement of the noise figure as a function of the source reflection coefficient, $F(\Gamma_S)$ [16]. Different input states were achieved by having a noise diode followed by a tuner, as shown in Figure 4.8b. The tuner was used to vary the source reflection coefficient, Γ_1, and for each value of Γ_1, the output power was measured with the diode in the "on" state (T_h) and in the "off" state ($T_c = T_{ambient}$). Note that the source reflection coefficient, Γ_1, must be the same for both hot and cold input sources. Let the set of reflection coefficients at reference plane 1 (at the tuner output) be Γ_i. Then for each Γ_i, the output power is measured with the diode turned on ($T_{h,i}$) and with the diode off ($T_{c,i}$), yielding the two measured power densities $p_{h,i}$ and $p_{c,i}$. The measured noise figure for this value of the reflection coefficient is computed by the Y-factor method,

$$F_i = 1 + \frac{T_{h,i} - Y_i T_{c,i}}{(Y_i - 1)T_0} \tag{4.30}$$

with $Y_i = p_{h,i}/p_{c,i}$. The noise parameters are then determined by a least-squares fit, minimizing the value of χ^2,

$$\chi^2 = \sum \frac{1}{u_i^2}[F_i - F(\Gamma_i)]^2 \tag{4.31}$$

where u_i is the uncertainty in the measurement of F_i, and $F(\Gamma_i)$ is the fitting function evaluated at Γ_i. If we use the parameterization of Eq. (4.13), then

$$F = F_{min} + \frac{R_n}{G_S}|Y_{opt} - Y_S|^2 \tag{4.32}$$

and the values for the noise parameters F_{min}, R_n, Y_{opt} are those that minimize Eq. (4.31), with $F(\Gamma_i)$ replaced by $F(Y_i, G_i)$. Often the uncertainties u_i are not known, or they are all assumed to be equal. In that case, the function to be minimized is just $\sum[F_i - F(\Gamma_i)]^2$. Any parameterization of the noise figure can be used; there is nothing special about Eq. (4.13). If a form with different noise parameters is used, then the values of that set of parameters are determined in the same manner, by minimizing Eq. (4.31). The choice of the set of input terminations will be discussed below.

In using the configuration of Figure 4.8b, it is important to note that hot and cold input noise temperatures are those at the input to the amplifier, i.e. the output of the tuner. Therefore, one must correct for the effect of the tuner on the diode noise temperature [26]. If the tuner is composed solely of passive components, it can be treated simply as an attenuator, using Eq. (1.28). If the cold noise temperature is the ambient temperature, T_{amb}, and if the tuner is also at ambient temperature, then the correction is not needed for $T_{c,i}$, but it is needed for $T_{h,i}$. If the tuner contains any active elements, such as PIN diode switches, then for careful measurements it is also necessary to correct for any extra noise power added by the active elements. In that case, it is necessary to measure the noise temperature(s) at the output of the tuner.

4.5.3 Fit to Noise-Temperature or Power Parameterization

Rather than measure the noise figure for a set of different input impedances or reflection coefficients, one can instead measure the output noise temperature or noise power for a set of input terminations and then perform a fit to a parameterization of the output noise temperature or power [17, 29]. This is the method that is most commonly used today. The advantage of this method is that it requires only one measurement with a nonambient input termination (or none, if the gain is measured by other means [30]), and it does not require a tuner (although a tuner can be used).

One starts with a parameterization of the noise temperature, such as

$$T_e(\Gamma_S) = T_{min} + \frac{4R_n T_0}{Z_0} \frac{|\Gamma_{opt} - \Gamma_S|^2}{|1 + \Gamma_{opt}|^2 \left(1 - |\Gamma_S|^2\right)} \tag{4.33}$$

or some equivalent form. In terms of T_e, the output noise temperature of the amplifier is given by

$$T_2 = G(\Gamma_1)[T_1 + T_e(\Gamma_1)] \tag{4.34}$$

where both the available gain and T_e depend on the reflection coefficient of the input termination, Γ_1. For the input termination T_i, Γ_i, the output temperature $T_2(T_i, \Gamma_i)$ in terms of the noise parameters is given by

$$T_2(T_i, \Gamma_i) = G(\Gamma_i)\left[T_i + T_{min} + \frac{4R_n T_0}{Z_0} \frac{|\Gamma_{opt} - \Gamma_i|^2}{|1 + \Gamma_{opt}|^2 \left(1 - |\Gamma_i|^2\right)}\right] \tag{4.35a}$$

in the IEEE representation, or

$$\begin{aligned} T_{2,i}(T_i, \Gamma_i) = \frac{|S_{21}|^2}{\left(1 - |\Gamma_{2,i}|^2\right)} &\left\{ \frac{\left(1 - |\Gamma_{1,i}|^2\right)}{|1 - \Gamma_{1,i} S_{11}|^2} T_{1,i} \right. \\ &\left. + \left|\frac{\Gamma_{1,i}}{1 - \Gamma_{1,i} S_{11}}\right|^2 X_1 + X_2 + 2Re\left[\frac{\Gamma_{1,i} X_{12}}{1 - \Gamma_{1,i} S_{11}}\right] \right\} \end{aligned} \tag{4.35b}$$

for the wave representation.

If the measured output noise temperature for the input termination i is denoted as $T_{2,i}^{meas}$, the function to be minimized is

$$\chi^2 = \sum \frac{1}{u_i^2} \left[T_{2,1}^{meas} - T_2(T_i \Gamma_i)\right]^2 \tag{4.36}$$

where u_i is the uncertainty in $T_{2,i}^{meas}$. As in Section 4.5.2, if the uncertainties are unknown or are all assumed to be equal, all the u_i are set equal to unity in Eq. (4.36). If, instead of Eq. (4.33), one preferred the wave representation of Eq. (4.19), then Eq. (4.35a) would be used in place of Eq. (4.35b), and the function to be minimized would still be Eq. (4.36).

Because of Eq. (4.35), the fitting function $T_2(T_i, \Gamma_i)$ includes the available gain $G(\Gamma_i)$, and there are two basic options for treating it. From Eq. (1.23), the available gain is given by

$$G(\Gamma_i) = \frac{|S_{21}|^2 \left(1 - |\Gamma_i|^2\right)}{|1 - \Gamma_i S_{11}|^2 \left(1 - |\Gamma_{2,i}|^2\right)} \tag{4.37}$$

where Γ_2 also depends on Γ_i, through

$$\Gamma_{2,i} = S_{22} + \frac{S_{12}S_{21}\Gamma_i}{(1 - \Gamma_i S_{11})} \tag{4.38}$$

One option is to measure S_{11} and the set of $\Gamma_{2,i}$ and to treat $|S_{21}|^2$ as an unknown parameter to be determined by the fit. In that case there would be five fitting parameters: T_{min}, R_n, complex Γ_{opt}, and $|S_{21}|^2$ (Instead of measuring the set of $\Gamma_{2,i}$, one could instead measure S_{22}, S_{21}, and S_{12}, and use Eq. (4.38) to compute the $\Gamma_{2,i}$. Direct measurement of the $\Gamma_{2,\mathrm{i}}$ results in smaller uncertainties, however.)

Rather than fit for $|S_{21}|^2$, one can use the value obtained in the VNA measurement of the amplifier's S parameters [30]. This strategy has the advantage that all the input terminations can be at ambient temperature, and no hot (or cryogenic) input termination is needed. This approach is sometimes referred to as "cold-only" measurements. Even if one determines $|S_{21}|^2$ from the fit to the noise measurements, it should be compared to the value from the VNA measurement, as a check.

The description in terms of output noise temperature, as in Eqs. (4.33)–(4.36), is natural for noise measurements on a radiometer. Most commercial instrumentation, however, measures delivered power, and the discussion is often framed in those terms. Since the noise temperature is proportional to the available noise-power spectral density, the conversion between noise temperature and delivered noise power is straightforward, and is accomplished simply by introducing the mismatch factor of Eq. (1.19),

$$p_2^{del} = M_2(\Gamma_i)k_B T_2 \tag{4.39}$$

with T_2 either the measured T_2 (for the measured delivered power) or the parameterization of Eq. (4.35) for the parameterization of $p_2^{del}(\Gamma_i)$. The noise parameters are then determined by minimizing

$$\chi^2 = \sum \frac{1}{u_i^2}\left[p_{2,i}^{meas} - p_2(T_i, \Gamma_i)\right]^2 \tag{4.40}$$

where $p_2(T_i, \Gamma_i)$ is the parameterization of the delivered power obtained from Eq. (4.39) and the parameterization of the output noise temperature that one chooses to use, e.g. Eq. (4.35). Sample results of noise-parameter measurements will be given below, when we discuss uncertainties in Section 4.7.

4.5.4 Possible Variations When Using the Wave Formulation

The wave representation of noise parameters, discussed above in Section 4.4.2, lends itself to direct measurement of certain noise parameters. We already mentioned the possibility of measuring X_{12} directly by combining the noise from input and output ports in a correlator. In addition, the physical identification of the parameter X_1 as the intrinsic noise emanating from the amplifier input suggests a strategy for its direct measurement [13]. If a perfectly matched load is connected to the amplifier's output port, as in Figure 4.6, then the noise temperature measured at the input port should be equal to X_1 (This is often referred to as the reverse noise temperature, T_{rev}, for obvious reasons.) In practice, of course, the matched load will have some small, nonzero reflection coefficient, resulting in some of the noise from the amplifier's output port being reflected back into the amplifier and contributing to the noise temperature at the input. The full form for the noise temperature at the input port, plane 1 in Figure 4.6, is

$$T_1 = \frac{1}{(1 - |\Gamma_1|^2)}[t_S + X_1 + t_2 + t_{12}] \tag{4.41}$$

where

$$t_S = \frac{|S_{12}|^2 (1 - |\Gamma_S|^2)}{|1 - \Gamma_S S_{22}|^2} T_S, \; t_2 = \left|\frac{S_{12}S_{21}\Gamma_S}{1 - \Gamma_S S_{22}}\right|^2 X_2, \; t_{12} = 2Re\left[\frac{S_{12}S_{21}\Gamma_S}{(1 - \Gamma_S S_{22})} X_{12}^*\right] \tag{4.42}$$

and where T_S is the noise temperature of the input termination.

Although X_1 is not the only contribution to T_1, it is typically the dominant one, since the others are suppressed by some combination of Γ_S or S_{12} (or both). Thus, performing a reverse measurement and including it in the results to be fitted can be an effective way to pin down the value of X_1. If such a direct measurement of T_1 is performed, Eq. (4.41) is included in the set of measurements to be fit. Alternatively, a reverse measurement can be used as a check. To use it as a check, one would determine the noise parameters by some standard procedure, fitting to a series of forward measurements, and would then use those measured noise parameters to predict the noise temperature T_1 in Figure 4.6, and compare this prediction to the result of the reverse measurement. This is discussed further in Chapter 6.1.

4.5.5 Choice of Input Terminations

All the prevailing methods for measuring amplifier noise parameters rely on an assortment of input states and a fit to the resulting measured outputs. Considerable effort has been expended to develop ways to choose an optimal set of input states, which would result in the "best" results with the least effort. In order to determine the four noise parameters plus a gain parameter, measurements with at least five different input terminations are required (four if the gain parameter is determined from VNA measurements). In choosing the input impedances, it is necessary to avoid certain singular configurations [31], and beyond that, one can try to optimize the set of input states. Optimization methods often require some initial knowledge of the amplifier's properties or employ an iterative method [32]. An alternate approach is to use a set of input states whose reflection coefficients provide a good coverage of the complex plane and can be used over a wide frequency range and for virtually any amplifier.

Clearly it is desirable to use more than the absolute minimum number of measurements, in order to reduce the effect of small errors in the measurements and to ensure a more robust measurement method. It is also intuitive that more measurements will be better, but that there is a point of diminishing returns, where the incremental improvement in the results does not justify the time and effort of the additional measurements. In practice, it appears that between 7 and 12 terminations provides sufficient redundancy without requiring excessive unnecessary effort. The reflection coefficients of the input terminations should be scattered throughout the Smith chart, ideally populating all four quadrants within the unit circle. In general, Γ_{opt} may not be known when the input terminations are chosen, and consequently it may not be possible to choose points based on their location relative to Γ_{opt}. And if one has a general idea of the value of Γ_{opt}, e.g. that it is near zero, the optimal strategy may seem a bit counter-intuitive. While one might at first think that the input terminations should cluster around Γ_{opt}, if its approximate value is known, distant points are also needed, especially to determine R_n [19]. To understand this, consider Figure 4.10, which uses Eq. (4.33) to plot T_e as a function of the complex Γ_S. The distant points are in a much steeper region of the paraboloid in Figure 4.10. Since R_n determines the slope of the paraboloid, the distant points are much more effective in determining its value.

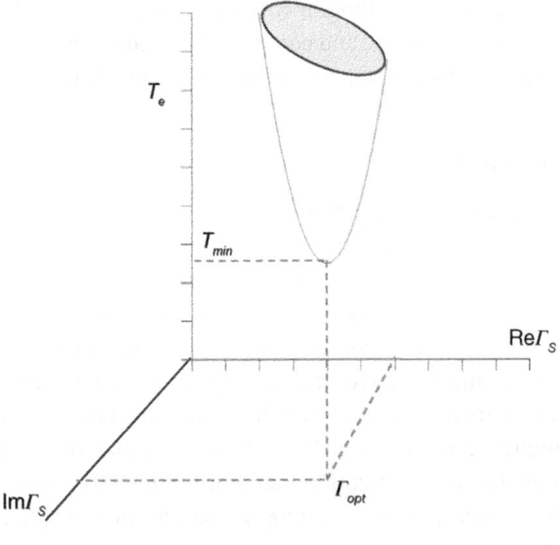

Figure 4.10 General shape of $T_e(\Gamma_S)$, from Eq. (4.33).

In choosing the specific pattern of the input impedances or reflection coefficients, the strategy options are quite different, depending on whether the measurement setup employs a tuner (as in Figure 4.8b or c), or a set of discrete terminations (as in Figure 4.8a). If a tuner is used, it is possible to choose any input impedance, subject to the tuner limitations, and one can attempt to optimize the input set. One can even employ a two-step process [32], in which a basic, reduced set of input states is used to determine the approximate location of Γ_{opt}, and this information is then used to choose additional source impedances to better determine the noise parameters. If a tuner is not used, the set of discrete impedances should be scattered across the complex Γ_S plane, bearing in mind that they will move as the frequency changes. A tuner can also be used to generate a predetermined set of input states, which are characterized ahead of time. A major reduction of measurement time is then achieved by sweeping frequency for each input termination, rather than stepping through all the input states for each frequency [33].

If one wishes to minimize the number of measurements, [34] and [35] have developed a prescription for choosing the four input terminations (assuming the gain parameter is determined by other measurements). They identified an optimal pattern for the input impedances or admittances and also investigated how much (or how little) the measurement uncertainties were affected by limiting the number of input states to only four. Even if one uses more than four input states, in order to reduce the uncertainties, it could be advantageous to include a base set of four that meet the criteria of [34] and [35].

In measurements at National Institute of Standards and Technology (NIST), a set of different discrete terminations was used, and a fixture was built to switch among them [27]. With discrete terminations, the reflection coefficients will change with frequency, and consequently the constellation of input states will vary with frequency. The important point is that the input reflection coefficients maintain a good coverage of the complex plane at each of the measurement frequencies. For this purpose, roughly 12 strategically chosen terminations seems to be sufficient. Assuming that one is trying to determine G_0 in addition to the four noise parameters, then at least five measurements are required. Adding additional input states improves the measurement uncertainty, but using more than about 12 pushes into the realm of diminishing returns.

Since G_0 is being determined, one of the input states is a matched load at non-ambient temperature; all others are at ambient temperature. The non-ambient temperature can be either hot or cold. In some circumstances, there can be some advantage to using a cold noise source as the non-ambient input termination [36], but usually a hot source is used because they are more readily available.

4.5.6 Commercial Systems, Source-Pull Measurements

There are a number of good commercial systems available for amplifier noise-parameter measurement. Most are based on some version of the fit to noise temperature or power measurements described in Section 4.5.3. The experimental setup is represented in Figure 4.7. The variable input can be achieved either by switching among different input sources or by use of a tuner, as indicated in Figure 4.8. Because the input or source is variable, such measurements are referred to as source-pull measurements.

In practice, one usually first measures the noise parameters of the receiver, as in Figure 4.11a. This is done by presenting the receiver with a series of input noise temperatures and reflection coefficients, and measuring the output noise power, resulting in a series of equations for the output noise power in terms of the receiver noise parameters. The receiver noise parameters are determined by a least-squares fit to that system of equations. In a similar manner, the configuration of Figure 4.11b is then used to measure the noise parameters of the DUT-receiver combination. Since the receiver noise parameters have already been measured, the DUT noise parameters can be extracted from those of the DUT-receiver combination. As in the method measuring noise factors, one must correct for the effect of the tuner on the input noise temperature.

In principle, the numerous measurements and multistep procedure is quite complicated. However, commercial systems designed for this purpose alleviate much of the difficulty. Care must still be taken, but the onus of detailed calculations and fits is relieved.

4.5.7 Frequency-Variation Method

An alternative method for producing different input reflection coefficients has been suggested [37] and implemented for noise-parameter measurements,

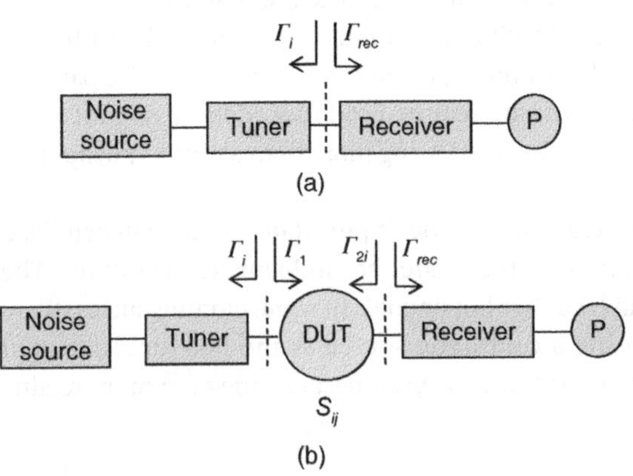

Figure 4.11 (a) Configuration for measuring the noise parameters of the receiver. (b) Configuration for measuring the noise parameters of the DUT once the receiver noise parameters are known.

especially those on transistors [38] and in cryogenic environments [39]. The idea is to use a mismatched input load with a length of transmission line at the amplifier input and to exploit the property of the line that the reflection coefficient at its output varies rapidly with frequency [40]. Thus by looking at different frequencies within a range Δf centered on the nominal measurement frequency f, one is looking at different source reflection coefficients. If (and only if) the noise parameters and S parameters of the DUT are constant across Δf, then the different frequency points within Δf provide different input terminations for the measurement of the noise parameters at the measurement frequency f. Because the delay line changes the phase but not the magnitude of the reflection coefficient, it is also useful to include a measurement with a matched load.

4.6 Uncertainty Analysis for Noise-Parameter Measurements

4.6.1 Simple Considerations

Because the determination of noise parameters typically involves least-squares (often nonlinear) fits to over-determined systems of equations, the full uncertainty analysis tends to be a complicated affair best addressed by Monte Carlo simulation. Before delving into the full analysis, however, it is instructive to consider the simpler case of the noise figure (or the effective input noise temperature) for the matched case, in order to gain some insight into features of the fits and measurements. Careful consideration of this simple case will also demonstrate that not everything is as simple as it may seem at first.

In Section 4.3, we saw that for the matched case the noise figure or effective input noise temperature can be measured by the Y-factor method, with the result that

$$T_e = \frac{T_h - YT_c}{Y - 1}, \qquad G = \frac{p_h - p_c}{k_B(T_h - T_c)} \tag{4.43}$$

and

$$F = 1 + \frac{T_e}{T_0} = 1 + \frac{T_h - YT_c}{(Y - 1)T_0} \tag{4.44}$$

where $Y = p_h/p_c$. In writing Eqs. (4.43) and (4.44), we have changed the notation of the output noise power densities from $N_{out,h}$ and $N_{out,c}$ to p_h and p_c, respectively. The earlier notation was used to distinguish between noise and signal powers and between input and output planes. We are now discussing only noise power densities at the output plane, and so we revert to the simpler notation.

To estimate the uncertainty in G and T_e, we use the usual law of propagation of uncertainty [41],

$$u^2(y) = \sum_{i=1}^{N} \left(\frac{\partial f}{\partial x_i}\right)^2 u^2(x_i) + 2\sum_{i=1}^{N-1}\sum_{j=i+1}^{N} \frac{\partial f}{\partial x_i}\frac{\partial f}{\partial x_j} u(x_i, x_j) \tag{4.45}$$

where the x_i are the measured variables on which $y = f(x_i)$ depends, $u(x_i)$ are the uncertainties in those measured quantities, and $u(x_i,x_j)$ is the covariance of x_i and x_j. It is usually a good approximation that there are no significant correlations among the errors in the measurements of T_h, T_c, and Y. In determining Y, however, there may be significant correlations in the errors in the output powers p_h and p_c. Thus, we write

$$u^2(T_e) \approx \frac{1}{(Y-1)^2}[u^2(T_h) + Y^2 u^2(T_c)] + \left[T_e^2 + \frac{T_c^2}{(Y-1)^2}\right] u^2(Y) \tag{4.46}$$

with

$$\frac{u^2(Y)}{Y^2} = \frac{u^2(p_h)}{p_h^2} + \frac{u^2(p_c)}{p_c^2} - \frac{2u(p_h, p_c)}{p_h p_c} \tag{4.47}$$

where $u(p_h, p_c)$ is the covariance of p_h and p_c. The uncertainty in the gain is given by

$$u^2(G) = \left(\frac{1}{k_B(T_h - T_c)}\right)^2 \{u^2(p_h) - 2u(p_h, p_c) + u^2(p_c) + k_B^2 G^2 [u^2(T_h) + u^2(T_c)]\} \tag{4.48}$$

The uncertainty in the gain is relatively simple in most cases. It is common to use a passive, ambient-temperature load (often a noise diode in its off state) as the cold source and a noise diode in its on state as the hot source. In that case, $u(T_h)$ is much larger than $u(T_c)$. Also, $k_B G u(T_h)$ is usually considerably larger than either $u(p_h)$ or $u(p_c)$, and furthermore, the covariance $u(p_h,p_c)$ is usually positive, further reducing the contribution of the output power measurements to the uncertainty in G. The result is that the fractional uncertainty in the gain can often be approximated by

$$\frac{u(G)}{G} \approx \frac{u(T_h)}{(T_h - T_c)} \tag{4.49}$$

The fractional uncertainty in the noise temperature of a commercial diode noise source is usually on the order of 2%, and so for a diode noise source with a noise temperature of 1500 or 9000 K, and a cold source at ambient temperature, the fractional uncertainty in the gain would also be in the neighborhood of 2%. For a diode noise source that is calibrated at a national standards laboratory, that fractional uncertainty could be reduced by a factor of about 2.

The uncertainty in the effective input noise temperature is somewhat more complicated, even in the simplified, matched case, but simplifying approximations can be made. Equations (4.46) and (4.47) are a bit unwieldy because common measurement parameters span a wide range of values for the variables appearing in them. However, matters become more tractable if we consider numeric values of common measurement configurations.

Consider first the terms in the first bracket in Eq. (4.46). The hot noise source is typically a commercial diode noise source with a noise temperature near either 1500 or 9000 K (the commonly available values), and the uncertainty in that noise temperature is roughly 2%. A rather tedious arithmetic exercise then shows that for either value of T_h, $u(T_h) \gg Yu(T_c)$ and the second term in the first bracket in Eq. (4.46) can be neglected.

For the $u^2(Y)$ contribution in Eq. (4.46), the sensitivity coefficient in the square brackets can be sizeable, particularly for large T_e; however, $u(Y)$ can usually be kept small. Any multiplicative errors in measuring the output power density will cancel in forming the ratio Y, and any additive errors can be eliminated by subtracting the reading with zero input power, i.e. using $Y = (p_h - p_0)/(p_c - p_0)$, where p_0 is the reading with no input signal. Such a ratio of differences is a very robust measurement; the only "major" residual errors would be due to instability (drift) or nonlinearity. Thus, the $u(Y)$ term in Eq. (4.46) can usually be made negligible, and we are left with

$$u(T_e) \approx \frac{u(T_h)}{(Y-1)} \tag{4.50}$$

As noted above, $u(T_h)$ is roughly $0.02\times\ T_h$ for most commercial diode noise sources, and the value of Y will depend on T_e of the DUT, through $Y \equiv p_h/p_c = (T_h + T_e)/(T_c + T_e)$. Thus, for a LNA with very small T_e, the uncertainty will be roughly 6 K for a T_h of either 1500 or 9000 K. As T_e increases, Y decreases, and the uncertainty increases. For example, if $T_e = 500$ K, the uncertainty will be around 16.5 K if $T_h = 9000$ K, and it will be around 20 K if $T_h = 1500$ K.

The expressions of Eqs. (4.46) and (4.48) assume that Eq. (4.43) is valid. We must also consider errors arising from effects that were ignored in writing Eq. (4.43). In particular, the assumption that the reflection coefficients of the hot and cold input sources both are zero (or even the same) is obviously not exactly correct. The additional error due to nonzero reflection coefficients of the input terminations will depend on the noise parameters of the DUT; it can be expected to be 1% or greater, depending also on the magnitude of the reflection coefficient.

The approximate results of this section were obtained for a simplified case in order to develop some intuition for the approximate magnitude to expect for the uncertainties. They rely on several approximations that are "usually" or "typically" valid, but these approximations should be checked in any specific application.

Common practices such as the use of an attenuator on the amplifier output or on the input of the power measurement system can degrade the uncertainties in the output power measurements and could invalidate some of the approximations. This is particularly true for on-wafer applications in Chapter 5, where the probe properties intrude.

4.6.2 Full Analysis

Contributions to the standard uncertainty can be divided into two groups [41]. Generally, type-A uncertainties are those that are determined by statistical means, and type-B are all others. The standard or combined uncertainty (u_c) in a quantity is the root sum of squares (RSS) of the type-A and type-B uncertainties,

$$u_c = \sqrt{u_A^2 + u_B^2} \tag{4.51}$$

Noise parameters are computed by performing a least-squares fit to an over-determined system of equations obtained by measuring the output noise temperature (or power) for each of a number of different input terminations connected to the amplifier or transistor under test. The type-A uncertainties can be computed from the covariance matrix of the fitted parameters. If the wave representation of Eq. (4.19) is used,

$$u_A(X_i) = \sqrt{V_{ii}(X)} \tag{4.52}$$

where X_i represents any of the five fitting parameters ($X_1, X_2, ReX_{12}, ImX_{12}$, and G_0). The covariance matrix obtained from the fit will be for the noise parameters used in the fit, e.g. the IEEE set ($T_{min}, t, \Gamma_{opt}, G_0$) or the wave-representation parameters (X_1, X_2, X_{12}, G_0). To obtain the type-A uncertainties in the other (or some other) representation, the covariance matrix in the desired representation is computed by using the Jacobian matrix for the transformation between the two representations. For example, if the fit was performed for the wave representation X-parameters, and if we use I_i to represent the IEEE parameters, and use V_{ij}(IEEE) and $V_{ij}(X)$ to represent the covariance matrix in the two representations, then the type-A uncertainties in the IEEE parameters are given by

$$u_A(I_i) = \sqrt{V_{ii}(IEEE)} \tag{4.53}$$

$$V_{ij}(IEEE) = \sum_{i',j'=1}^{5} \frac{\partial I_i}{\partial X_{i'}} \frac{\partial I_j}{\partial X_{j'}} V_{i'j'}(X)$$

where the elements of the Jacobian matrix ($\partial I_i/\partial X_{i'}$) are computed from Eqs. (4.22) and (4.23). The computation is straightforward but tedious, and the results are lengthy and unenlightening. They can be found in [42]. As a practical matter, for precision measurements it is usually possible to perform enough redundant measurements to render type A uncertainties negligible compared to type B – assuming measurement time is not a constraint.

The type B uncertainties are less straightforward and require more effort. We assume that the uncertainties in the underlying or input quantities, such as reflection coefficients, measured noise temperatures, etc. are known or can be estimated; but the problem of propagating these underlying uncertainties to compute uncertainties in the output noise parameters generally does not admit a simple analytical solution. We do not have an equation for the noise parameters in terms of the underlying measured quantities, and therefore we cannot simply apply Eq. (4.45). Therefore, a Monte Carlo approach is used for the type B uncertainties [19, 42–48]. (Within a particular noise-parameter analysis, an analytic form for the uncertainty due to errors in input reflection coefficients has been developed and studied [49].)

The Monte Carlo uncertainty evaluation is based on the generation of simulated results for the measurements that were actually performed (see, e.g. [50]). These simulated measurement results are analyzed in the same way as are the real data, yielding simulated values for the noise parameters. This process is repeated a large number of times, and the distribution of values obtained for each noise parameter is used to compute its uncertainty. One full simulation of all the measurements that were actually performed will be referred to as a "set" of measurements, and the number of simulated measurement sets will be denoted N_{sim}, which is chosen to be large enough that the computed uncertainties are approximately independent of N_{sim}. A value of $N_{sim} = 10\,000$ is usually more than sufficient, but for poorly matched transistors, it is occasionally necessary to use larger values. In any case, one should use a value large enough that any further increase in N_{sim} changes the uncertainties by at most 10% of their values. In most cases, the uncertainties should be within a few percent of their asymptotic values.

Each simulated measurement is performed by choosing a number randomly from the probability distribution for that physical quantity. The distribution is constrained to have an average value equal to the true value and a standard deviation equal to the standard uncertainty for that quantity. The actual measurement results are used as the true values. Normal distributions are usually assumed for all quantities unless additional information about the distribution is available. For example, in the results shown below, a rectangular distribution was used for the ambient temperature, because it more realistically models the effect of a thermostatically controlled laboratory temperature (Using a normal distribution for the ambient temperature makes virtually no difference in practice.) For each measurement set, one first simulates measurement of the input parameters; these include the DUT S-parameters, the reflection coefficients of all input terminations, the noise temperatures of all terminations, and the output reflection coefficients, $\Gamma_{2,i}$ in Figure 4.7 (assuming that the $\Gamma_{2,i}$ are measured rather than computed). To simulate the measurement of the output noise temperatures, one must compute the true values from the equation giving the output noise power or noise temperature in terms of the noise parameters, e.g. Eqs. (4.35a) or (4.35b). In computing the true output noise temperature, the true (i.e. measured) values are used for all quantities on the right-hand side.

Because the input variables include sets of variables that are all measured in the same manner, on the same equipment, correlated errors can and do occur, and the correlations can cause significant effects in the noise-parameter uncertainties. Therefore, correlated errors should be included in the simulations. At the least, the correlations that should be included are those among all the measured reflection coefficients and the S-parameters, among all the output noise temperatures, and among the temperatures of the ambient-temperature terminations.

Each set of simulated noise-temperature measurements is analyzed in the same manner as a set of real measurements: one performs a least-squares fit to the measurement results and obtains a set of the noise parameters. The particular set of noise parameters obtained is the set used in the equation for the output noise power or noise temperature. If one fits to Eq. (4.35a), the fit determines the noise parameters in the IEEE representation (T_{min}, t, Γ_{opt}, G_0), and if one fits to Eq. (4.35b), the fit determines the noise parameters in the wave representation (X_1, X_2, X_{12}, G_0). For each simulated measurement set, the noise parameters in the set that was not used in the fit are then computed from those that were used in the fit plus the simulated measurement results for the amplifier's S-parameters. This is done for each of the N_{sim} sets of simulated measurements. The average and standard deviation of the set of simulated measured values for each parameter (X and IEEE) are computed. The (type-B) uncertainty of a parameter is then computed by combining the standard deviation in quadrature with the difference between the average and the true value. This is just the root mean square error (RMSE) of the sample,

$$u_B(y) = RMSE(y) = \sqrt{Var(y) + (\overline{y} - y_{true})^2} \tag{4.54}$$

where y represents any of the noise parameters, $\overline{y}$ is the average of the sample of simulated results for y, and $Var(y)$ is the variance of the sample of simulated results for y. For the complex quantities X_{12} and Γ_{opt}, statistics are done on the real and imaginary parts separately. The fact that $\overline{y}$ is not equal to the true value may be unsettling at first, but such are the vagaries of nonlinear functions.

4.6.3 Input Uncertainties

The simulations require as input not just the true (i.e. measured) values of all parameters, but also the standard uncertainties in the parameters that are directly measured, i.e. the amplifier S-parameters, the reflection coefficients of all terminations, the noise temperatures of all terminations, the output reflection coefficients $\Gamma_{2,i}$ (if they are measured), and the output noise temperatures. These input uncertainties are rather straightforward for measurements on packaged amplifiers, except for estimating correlated errors.

Scattering parameters and reflection coefficients are typically measured on a commercial VNA, and information about the uncertainties is usually available from the manufacturer. The uncertainty will depend on the connector type, frequency, magnitude of the reflection coefficient being measured, and method used to calibrate the VNA. Values of $u(|\Gamma|)$ in the range 0.001–0.005 are not unreasonable (larger for higher frequencies or on-wafer). Unfortunately, information regarding correlations in the measurements is not always available. The effect of connector nonrepeatability can either be included in the S-parameter and reflection coefficient uncertainties, or it can be introduced separately, but it should be included. Even for good connectors at low frequencies (the low-GHz range), it is difficult to achieve connector repeatability better than 0.001, and it is easy to do much worse. Since all reflection coefficients are usually measured with the same VNA calibration, and since the VNA calibration and standards are a significant source of any measurement error, correlations in the VNA errors can be expected to be significant. In [45, 47], the correlation coefficient was taken to be $\rho \approx 0.86$ or $\rho \approx 0.94$, depending on the magnitude of the reflection coefficient.

For the uncertainty in the ambient temperature, we must be careful to distinguish between the ambient or room temperature and the temperature of the ambient standard. The uncertainty in the ambient standard is treated in the uncertainty of the output noise temperature or power, as in Chapter 3. The uncertainty in the "ambient-temperature" input terminations are treated separately. They are often passive loads with no extra thermal control or monitoring, in which case the uncertainty in their noise temperature is the uncertainty in the room temperature (assuming that there is no nearby heat source). The uncertainty can be improved by encasing the terminations and controlling or measuring the temperature in the case. Excessive effort is not usually necessary, since the errors in the ambient noise temperatures usually have negligible effect on the overall uncertainties, unless they get very large (about 2 K or larger). One point to note is that if active (pin diode) switches are used to switch among the ambient terminations (as in electronic calibrations units), they can add additional noise, and the noise temperature of the terminations should be measured rather than assumed to be that of a passive termination at the ambient physical temperature.

Uncertainties in the measured output noise powers or noise temperatures will, of course, depend on the particular method or system used to measure them. In principle, the uncertainty will depend on the power level, the relevant reflection coefficients, the frequency, and the standards used to calibrate the system. In practice, it may be possible (and it is certainly desirable) to use a simple algorithm or equation to compute the uncertainty for a given output noise temperature. Depending on the measurement method used, there will be large correlations in the errors in the output noise measurements for the different terminations, particularly if all the reflection coefficients are measured with the same VNA calibration. As will be seen below, the correlation in the errors in the output noise measurements reduces the errors in the noise parameters. Therefore, underestimating this correlation leads to larger, more conservative estimates of the uncertainty.

4.6.4 General Features and Sample Results

The magnitudes of the uncertainties in the noise parameters depend on a host of factors, including (but not limited to) the input reflection coefficients and their uncertainties, the uncertainties in the measurement of the output noise temperatures or powers, and the nonambient input noise temperature and its uncertainty, as well as the properties of the DUT itself. There is no simple expression for the noise-parameter uncertainties, but some general approximate features can be inferred from the results for a variety of DUTs and input uncertainties [42].

For the IEEE parameters, assuming the common situation of one hot input source and multiple ambient-temperature sources, the uncertainties in the gain parameter and T_{min} (or F_{min}) are dominated by the uncertainty in the hot input noise temperature. An uncertainty of 0.1 dB in the hot input noise temperature leads to an 0.1 dB uncertainty in G_0 and F_{min}. Not surprisingly, uncertainties in Γ_{opt} are dominated by the uncertainties in the input reflection coefficients. The uncertainty in the real or imaginary part of Γ_{opt} is about three or four times as large as the uncertainty in the real or imaginary parts of the input reflection coefficients (for the case of 13 input terminations). The uncertainty in the parameter t (or R_n) is sensitive to almost everything; there is no one dominant factor. The uncertainty in the ambient temperature is not a dominant factor in the noise-parameter uncertainties, primarily because T_{amb} is known so much better than the hot input noise temperature. However, the ambient (room) temperature could affect the noise temperature of the input hot source or the properties of the DUT, assuming that they are at room temperature.

The connector variability and the uncertainty in the ambient temperature have little effect on the uncertainties in any of the output variables, except possibly in extreme cases. Correlations among the underlying uncertainties increase the output uncertainties in some cases and decrease them in others; the most dramatic effects of correlations are reductions in the uncertainty in the gain and Γ_{opt} for correlated errors in measuring the output noise temperature [45]. Reference [45] also shows the dependence of the uncertainty on one of the input uncertainties when all other input uncertainties are zero.

To show the approximate uncertainties that are achievable, results from measurements on a LNA from 8 to 12 GHz are shown in Figure 4.12 and 4.13a–f. The measurements were performed in the course of testing a variable-termination unit (VTU) [27], which can be used to automatically switch among a series of input terminations, greatly reducing the time required for the measurements. For these measurements, 11 different input terminations were used in the forward configuration, of which one was hot (between 1000 and 1100 K, depending on frequency), one cold (about 100–110 K), and the remainder at room temperature (about 296.15 K). In addition, one reverse measurement was done, with a matched load on the output of the LNA.

Results for the X parameters are shown in Figure 4.12. The quantity X_2 is the equivalent input noise temperature for a matched (reflectionless) input termination, often called the 50 Ω noise temperature. There are two sets of uncertainties plotted for X_2, to show the effect of an output attenuator on the measurements. The actual measurements were performed with a 20 dB attenuator on the amplifier's output (for the forward configuration), in order to keep the output noise temperature in the measurement range of the radiometer used [51]. Consequently, the

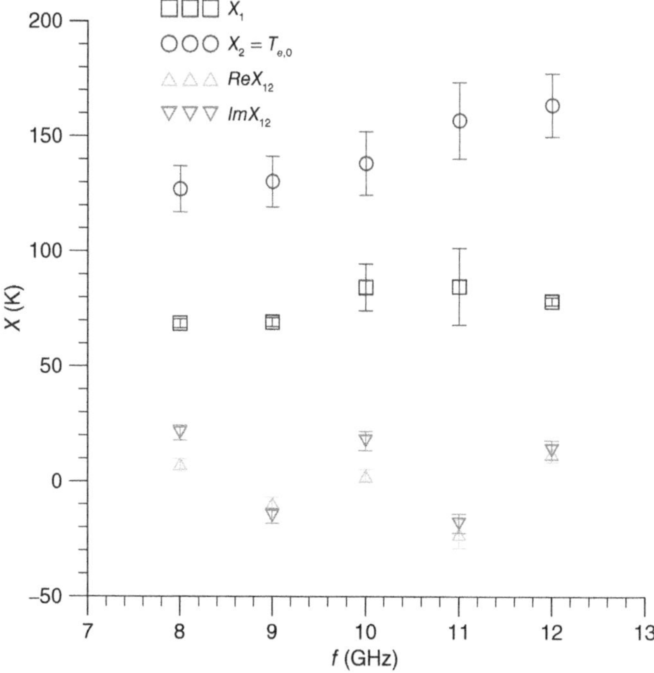

Figure 4.12 Measurement results and uncertainties for wave-representation noise parameters of an LNA. Source: From Randa [42]/U.S. Department of Commerce/Public Domian.

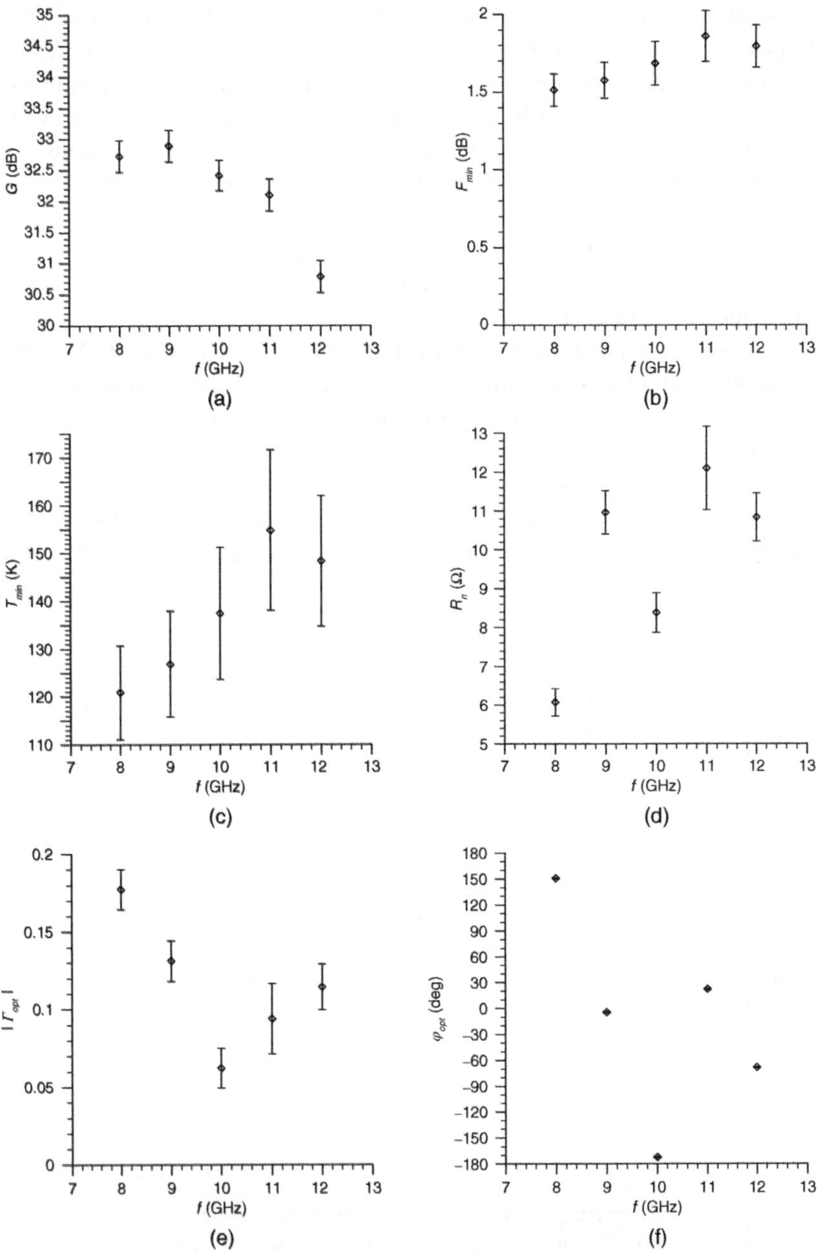

Figure 4.13 Measurement results and uncertainties for IEEE noise parameters of an LNA. (a) Measured reduced gain G_0. (b) Measured F_{min}. (c) Measured T_{min}. (d) Measured R_n. (e) Measured Γ_{opt}. (f) Measured ϕ_{opt}. Source: From Randa [42]/U.S. Department of Commerce/Public Domian.

noise at the radiometer/DUT measurement plane is weighted 99 to 1 in favor of the attenuator over the amplifier, and the uncertainty suffers significantly. The larger uncertainties on X_2 in Figure 4.12 correspond to the actual measurements, and the smaller uncertainties represent what could be achieved if no attenuator were needed.

In Figures 4.13a–f, we show the results for G_0 and the IEEE noise parameters. Because it is in common use, we have also shown the minimum noise figure in dB, which is related to T_{min} by $F_{min}(dB) = 10log_{10}(1 + T_{min}/T_0)$. As in Figure 4.12, the two sets of uncertainties correspond to the actual measurements (larger uncertainties) and the case with no output attenuator (smaller uncertainties). For the phase of Γ_{opt}, Figure 4.13f, we show only the actual uncertainties, since they are so small. As it does for the X parameters, the use of the attenuator causes a major increase in the uncertainties. The approximate values for the actual uncertainties are: 0.25 dB for G_0, 0.10–0.16 dB for F_{min}, 10–17 K for T_{min}, 6–12 Ω for R_n, 0.013– 0.023 for $|\Gamma_{\text{opt}}|$, and 0.8–1.0° for ϕ_{opt}. If an attenuator had not been necessary, these uncertainties would have been 0.05 dB for G_0, 0.03 dB for F_{min}, 2.5–3.5 K for T_{min}, 0.08–0.16 Ω for R_n, 0.003–0.007 for $|\Gamma_{\text{opt}}|$, and 0.2–0.5° for ϕ_{opt}.

4.7 Simulations and Strategies

There are many variations of the general strategies for measuring noise parameters, and the Monte Carlo techniques employed in the uncertainty analysis can also be used to compare the efficacy of the different strategies. Simulations have been used to study the effect using a cryogenic source vs. a hot source for the non-ambient temperature (for cases in which the gain parameter is determined by the noise measurements), of using multiple non-ambient input terminations, of increasing (or decreasing) the number of input terminations, and of one or more reverse measurements [36].

A limitation of drawing conclusions from simulation results is that strictly speaking, the results only apply to the specific case that was simulated, to that particular set of DUT noise parameters, S-parameters, reflection coefficients, etc. By looking at results for a range of different situations, it is possible to infer "typical" results or properties, but without examining all possible cases, it is not possible to rule out the possibility of exceptional cases that contravene the general properties that one might infer. Thus, the results should be regarded as general "rules of thumb," rather than some sort of rigorous result.

Some of the general results can be summarized as follows. The noise temperature of the commonly used hot input matched load should be as far from ambient as possible (while still keeping the DUT and output measurements system in their linear operating ranges). It is very helpful to use a cold (cryogenic) input matched load, particularly for low-noise devices. Substituting a cold load for the hot input load yields some improvements for many amplifiers, but use of a cold load *in addition to* the hot load yields larger improvements in the uncertainties for all the amplifiers considered. Inclusion of a reverse measurement helps, but it requires a different measurement configuration, and the improvement in uncertainties is not always dramatic. The practical recommendation that emerges is to add a cold (cryogenic) matched load to the usual set of input terminations; for low-noise devices it will reduce the uncertainty in T_{min} by a factor of about two.

The issue that probably has the most practical importance is the number of input terminations that should be used and where their reflection coefficients should be located in the complex plane. Using too few terminations leads to poorly determined noise parameters; using too many results in wasted time and effort. The absolute minimum number of input terminations required is five if the gain parameter is to be measured along with the noise parameters, and the five must include at least two different noise temperatures. If the gain parameter is determined from the S-parameter measurements, then the absolute minimum is four, all of which can have the same noise temperature. Those are mathematical properties that do not depend on simulations.

It is common practice to use more than the bare minimum number of input terminations, and the question that then arises is how many, and what reflection coefficients. We already discussed the choice of input reflection coefficients to some extent in Section 4.5.5. Some general rules of thumb can be inferred from simulation results (These all assume that nothing is known about Γ_{opt} in advance.) Roughly eight to 12 different input terminations (including the two matched loads at different noise temperatures) are needed in order to get good results. They should populate all four quadrants of the unit circle in the complex plane and should be scattered rather than bunched. Some (up to four) should be highly reflective, located near the edge of the unit circle. This assures that there will be larger distance from some measurement points to Γ_{opt}, thereby providing more sensitivity to the location of Γ_{opt} [19]. The common practice of including interior points (neither reflective nor matched) does indeed lead to better results. The output reflection coefficients should definitely be measured rather than computed by cascade. Additional (beyond four) reflective input terminations do not help and in fact may hurt – in some circumstances they can lead to an increase in the likelihood of obtaining unphysical results. For more details, the reader is referred to [36].

As discussed in Section 4.5.5, schemes to choose an optimal set of reflection coefficients typically require that Γ_{opt} be known in advance or be measured in a preliminary round of measurements. These results can be used in an iterative scheme, in which one first performs a quick, crude measurement of Γ_{opt} and then uses that knowledge to choose input reflection coefficients for a refined measurement. It is not clear, however, that such an approach yields improved uncertainties, and it does require more work. The optimal set of [35] optimizes number of measurements rather than uncertainty, but as noted in Section 4.5.5, it can certainly be included in the set of input terminations.

References

1 R. Adler, R.S. Engelbrecht, S.W. Harrison, H.A. Haus, M.T. Lebenbaum, and W.W. Mumford, "Description of the noise performance of amplifiers and receiving systems," *Proceedings of the IEEE*, **51**, no. 3, pp. 436–442 (March 1963).

2 H.T. Friis, "Noise figures of radio receivers," *Proceedings of the IRE*, **32**, no. 7, pp. 419–422 (1944).

3 H. Rothe and W. Dahlke, "Theory of noisy fourpoles," *Proceedings of the IRE*, **44**, pp. 811–818 (1956).

4 H. Hillbrand, and P.H. Russer, "An efficient method for computer aided noise analysis of linear amplifier networks," *IEEE Transactions on Circuits and Systems*, **23**, pp. 235–238 (1976).

5 J. Engberg and R. Larsen, *Noise Theory of Linear and Non-linear Circuits*, Wiley, New York, 1995.

6 S.A. Maas, *Noise in Linear and Nonlinear Circuits*, Artech House, Norwood, MA, 2005.

7 H.A. Haus, W.R. Atkinson, G.M. Branch, W.B. Davenport Jr., W.H. Harris, S.W. Harrison, W.W. McLeod, E.K. Stodola, and T.E. Talpey, "Representation of noise in linear twoports," *Proceedings of the IRE*, **48**, pp. 69–74 (1960).

8 H.A. Haus et al., "IRE standards on methods of measuring noise in linear twoports 1959," *Proceedings of the IRE*, **48**, pp. 60–68 (1960).

9 P. Penfield Jr,., "Wave representation of amplifier noise," *IRE Transactions on Circuit Theory*, **CT-9**, pp. 84–86 (March 1962).

10 G.F. Engen, "A new method of characterizing amplifier noise performance," *IEEE Transactions on Instrumentation and Measurement*, **IM-19**, no. 4, pp. 344–349 (November 1970).

11 R.P. Meys, "A wave approach to the noise properties of linear microwave devices," *IEEE Transactions on Microwave Theory and Techniques*, **MTT-26**, pp. 34–37 (January 1978).

12 R.P. Hecken, "Analysis of linear noisy two-ports using scattering waves," *IEEE Transactions on Microwave Theory and Techniques*, **MTT-29**, no. 10, pp. 997–1004 (October 1981).

13 D. Wait and G.R. Engen, "Application of radiometry to the accurate measurement of amplifier noise, *IEEE Transactions on Instrumentation and Measurement*, **40**, pp. 433–437 (April 1991).

14 S.W. Wedge and D.B. Rutledge, "Wave techniques for noise modeling and measurement," *IEEE Transactions on Microwave Theory and Techniques,* **40**, pp. 204–212 (1991).

15 S.W. Wedge and D.B. Rutledge, "Wave techniques for noise modeling and measurement," *IEEE Transactions on Microwave Theory and Techniques*, **40**, no. 11, pp. 2004–2012 (November 1992).

16 R.Q. Lane, "The determination of device noise parameters," *Proceedings of the IEEE*, **57** pp. 1461–1462 (August 1969).

17 V. Adamian and A. Uhlir, "A novel procedure for receiver noise characterization," *IEEE Transactions on Instrumentation and Measurement*, **IM-22**, pp. 181–182 (June 1973).

18 M.W. Pospieszalski, "On the measurement of noise parameters of microwave two-ports," *IEEE Transactions on Microwave Theory and Techniques*, **MTT-34**, no. 4, pp. 456–458 (April 1986).

19 A.C. Davidson, B.W. Leake, and E. Strid, "Accuracy improvements in microwave noise parameter measurements," *IEEE Transactions on Microwave Theory and Techniques*, **37**, no. 12, pp. 1973–1978 (December 1989).

20 A. Boudiaf and M. Laporte, "An accurate and repeatable technique for noise parameter measurements," *IEEE Transactions on Instrumentation and Measurement*, **42**, no. 2, pp. 532–537 (April 1993).

21 G. Martines and M. Sannino, "The determination of the noise, gain and scattering parameters of microwave transistors …", *IEEE Transactions on Microwave Theory and Techniques*, **42**, no. 7, pp. 1105–1113 (July 1994).

22 G.L. Williams, "Measuring amplifier noise on a noise source calibration radiometer," *IEEE Transactions on Instrumentation and Measurement*, **44**, no. 2, pp. 340–342 (April 1995).

23 D.F. Wait and J. Randa, "Amplifier noise measurements at NIST," *IEEE Transactions on Instrumentation and Measurement*, **46**, no. 2, pp. 482–485 (April 1997).

24 T. Werling, E. Bourdel, D. Pasquet, and A. Boudiaf, "Determination of wave noise sources using spectral parametric modeling," *IEEE Transactions on Microwave Theory and Techniques*, **45**, no. 12, pp. 2461–2467 (December 1997).

25 A. Lazaro, L. Pradell, and J.M. O'Callaghan, "FET noise-parameter determination using a novel technique based on 50-Ω noise-figure measurements," *IEEE Transactions on Microwave Theory and Techniques*, **47**, no. 3, pp. 315–324 (March 1999).

26 E. Strid, "Measurement of losses in noise-matching networks," *IEEE Transactions on Microwave Theory and Techniques*, **MTT-29**, no. 3, pp. 247–252 (March 1981).

27 D. Gu, D.K. Walker, and J. Randa, "Variable termination unit for noise-parameter measurement," *IEEE Transactions on Instrumentation and Measurement*, **58**, no. 4, pp. 1072–1077 (April 2009).

28 G.L. Williams, "A broad-band radiometer for calibration of mismatched noise sources," *IEEE Transactions on Instrumentation and Measurement*, **40**, no. 2, pp. 443–445 (April 1991).

29 V. Adamian and A. Uhler, "Simplified noise evaluation of microwave receivers. *IEEE Transactions on Instrumentation and Measurement*, **33**, pp. 136–140 (1984).

30 Keysight Application Note 5990–5800, "High-accuracy noise-figure measurements using the PNA-X series NA. https://www.keysight.com/zz/en/assets/7018-02539/application-notes/5990-5800.pdf (Accessed 09 July, 2022).

31 G. Caruso and M. Sannino, "Couputer-aided determination of microwave two-port noise parameters," *IEEE Transactions on Microwave Theory and Techniques*, **MTT-26**, no. 9, pp. 639–642 (September 1978).

32 S. Van den Bosch and L. Martens, "Improved impedance-pattern generation for automatic noise-parameter determination," *IEEE Transactions on Microwave Theory and Techniques*, **46**, no. 11, pp. 1673–1678 (November 1998).

33 G. Simpson, D. Ballo, J. Dunsmore, and A. Ganwani, "A new noise parameter measurement method results in more than 100× speed improvement and enhanced measurement accuracy," in *81st ARFTG Conference Digest*, Portland, OR, December 2008, pp. 119–127.

34 M. De Dominicis, F. Giannini, E. Limiti and G. Saggio, "A novel impedance pattern for fast noise measurements," *IEEE Transactions on Instrumentation and Measurement*, **51**, no. 3, pp. 560–564 (June 2002). doi: 10.1109/TIM.2002.1017728.

35 M. Himmelfarb and L. Belostotski, "On impedance-pattern selection for noise parameter measurement," *IEEE Transactions on Microwave Theory and Techniques*, **64**, no. 1, pp. 258–270 (January 2016). doi: 10.1109/TMTT.2015.2504500.

36 J. Randa, "Comparison of noise-parameter measurement strategies: simulation results for amplifiers," *84th ARFTG Microwave Measurement Conference*, Boulder, CO, 2014, pp. 1–8. doi: 10.1109/ARFTG.2014.7013402.

37 V.D. Larock and R.P. Meys, "Automatic noise temperature measurement through frequency variation," *IEEE Transactions on Microwave Theory and Techniques*, **30**, no. 8, pp. 1286–1288 (August 1982). doi: 10.1109/TMTT.1982.1131242.

38 R. Hu and T.-H. Sang, "On-wafer noise-parameter measurement using wide-band frequency variation method," *IEEE Transactions on Microwave Theory and Techniques*, **53**, no.7, pp. 2398–2402 (July 2005).

39 R. Hu and S. Weinreb, "A novel wide-band noise-parameter measurement method and its cryogenic application," *IEEE Transactions on Microwave Theory and Techniques*, **52**, no. 5, pp. 1498–1507 (May 2004).

40 D.M. Pozar, *Microwave Engineering*, 2, Chap. 2, Wiley, New York, U.S.A., 1998.

41 ISO. *ISO Guide to the Expression of Uncertainty in Measurement, International Organization for Standardization*, ISO, Geneva, Switzerland, 1993.

42 J. Randa, "Uncertainty analysis for NIST noise-parameter measurements, NIST Technical Note 1530 (March 2008). Available online: https://permanent.fdlp.gov/gpo12352/08-TN1530-NPUncerts.pdf (Accessed 20 March, 2022.)

43 J. Randa, and W. Wiatr, "Monte Carlo simulation of noise parameter uncertainties," in *IEE Proceedings-Science Measurement and Technology*, [online] (2002), https://tsapps.nist.gov/publication/get_pdf.cfm?pub_id=12549 (Accessed 17 March, 2022)

44 A. Boudiaf, M. Laporte, J. Dangla, and G. Vernet, "Accuracy improvements in two-port noise parameter extraction method," in *1992 IEEE MTT-S International Microwave Symposium Digest*, Albuquerque, NM, pp. 1569–1572 (1992).

45 J. Randa, "Noise-parameter uncertainties: a Monte Carlo simulation," *Journal of Research of the National Institute of Standards and Technology*, **107**, no. 5, pp. 431–444 (2002). https://dx.doi.org/10.6028/jres.107.037. Correction: *ibid*, **111**, p. 461 (November 2006). http://dx.doi.org/10.6028/jres.111.035.

46 T. Vähä-Heikkilä, M. Lahdes, M. Kantanen, and J. Tuovinen, "On-wafer noise-parameter measurements at W-band," *IEEE Transactions on Microwave Theory and Techniques*, **51**, no. 6, pp. 1621–1628 (June 2003).

47 J. Randa, "Uncertainty analysis for noise-parameter measurements at NIST," *IEEE Transactions on Instrumentation and Measurement*, **58**, no. 4, pp. 1146–1151 (April 2009).

48 L. Belostotski and J.W. Haslett, "Evaluation of tuner-based noise-parameter extraction methods for very low noise amplifiers," *IEEE Transactions on Microwave Theory and Techniques*, **58**, no. 1, pp. 236–250 (2010).

49 W. Wiatr and D.K. Walker, "Systematic errors of noise parameter determination caused by imperfect source impedance measurement," *IEEE Transactions on Instrumentation and Measurement*, **54**, no. 2, pp. 696–700 (April 2005).

50 W.H. Press, B.P. Flannery, S.A. Teukolsky, and W.T. Vetterling, *Numerical Recipes*, Chap. 14.5, Cambridge University Press, Cambridge, U.K., 1986.

51 C.A. Grosvenor, J. Randa, and R.L. Billinger, "Design and testing of NFRad—a new noise measurement system," NIST Technical Note 1518 (March 2000). https://tsapps.nist.gov/publication/get_pdf.cfm?pub_id=6264 (Accessed 17 March, 2022).

5

On-Wafer Noise Measurements

5.1 Introduction

In a sense, on-wafer noise measurements are just measurements through adapters, the adapters being the on-wafer probes with coaxial connectors. Thus, measuring noise parameters on-wafer is similar to measuring them in a coaxial environment – except harder. The additional difficulties are due to the probes and the open on-wafer environment as well as to the properties of the devices that are typically measured.

It is necessary to characterize the probes, of course. This involves use of on-wafer standards, which leads to larger uncertainties for the reflection coefficients of the input terminations, the device under test (DUT) S-parameters, and for T_{in} and T_{out} of the DUT. The probe losses also restrict the range of available reflection coefficients for the input terminations, assuming that they are off-wafer terminations connected to the DUT input through a probe. The probe is also subject to possible repeatability difficulties, due to potential contact problems and vibrations. And the open on-wafer environment raises the risk of exposure to stray radiation and to temperature gradients.

Complications due to the device depend on the type of device being measured. If measuring the noise parameters of an on-wafer amplifier, there are no additional device-related problems (assuming that it is well matched). However, the devices whose noise parameters are to be measured on-wafer are often transistors rather than amplifiers, and that can introduce additional difficulties. A transistor can be poorly matched, with large values of S_{11} and S_{22}, leading to larger corrections and therefore larger uncertainties. Also, a transistor is more likely than an amplifier to have a large value of $|\Gamma_{opt}|$, near the edge of the Smith chart, which is more difficult to measure accurately, particularly since the range of input reflection coefficients is restricted by the probe loss. Finally, transistors usually have smaller values of the minimum noise figure or effective input noise temperature, and smaller noise temperatures are more difficult to measure.

Precision Measurement of Microwave Thermal Noise, First Edition. James Randa.

5.2 On-Wafer Microwave Formalism

5.2.1 Traveling Waves vs. Pseudo Waves

In on-wafer measurements, some care needs to be taken with the definition and treatment of the waves in the on-wafer transmission lines. Because on-wafer transmission lines can exhibit significant loss, the equation for the power in terms of traveling waves (for a single mode) is no longer just $p = |a|^2 - |b|^2$, where a and b are the (appropriately normalized) amplitudes of the traveling waves in the two opposite directions on the transmission line. Instead, there is an additional cross term, $p = |a|^2 - |b|^2 + 2tan\zeta Im(ab^*)$, where ζ is the phase of the characteristic impedance of the line for that mode [1]. As a consequence, the expressions derived in Chapter 1 for quantities such as the mismatch factor, the efficiency, and the available power ratio in terms of the traveling waves need to be modified. The modified equations are derived in [2].

Use of the (rather cumbersome) expressions in terms of the traveling waves in lossy lines can be avoided, however. Marks and Williams [3] have suggested and developed use of what they call "pseudo-waves," which are linear combinations of the traveling waves, where the linear combination depends on the characteristic impedance and on a reference impedance of the user's choice. The transformation to pseudo-waves induces a corresponding transformation of reflection coefficients and S-parameters at the reference planes affected. The advantage of pseudo-waves is that if one chooses a real reference impedance (50 Ω, for example), then the expression for the delivered power reverts to the familiar lossless-line expression, $|a|^2(1 - |\Gamma|^2)$. Consequently, the expressions for ratios of powers, such as mismatch factors, efficiencies, and available-power ratios, also revert to the lossless-line forms of Chapter 1.

The price that one pays for using pseudo-waves is that the reflection coefficients and scattering parameters must also be transformed to the pseudo-wave forms for the chosen reference impedance. Fortunately, software ([4]) exists that enables the on-wafer calibration and measurement in terms of pseudo-wave quantities. One thus has a choice: one can use traditional traveling waves, in which case one must modify the equations involving powers (efficiency, mismatch factor, and available power ratio), or one can use pseudo-waves, in which case the old traditional forms for powers are used, but one must use the pseudo-wave reflection coefficients and S-parameters. This is discussed in considerable detail in [2]. In the following sections we assume the use of pseudo-waves with a real reference impedance of 50 Ω.

5.2.2 On-Wafer Reference Planes

In any vector-network-analyzer (VNA) measurements, it is important to be clear about the location of reference planes, and this is true *a fortiori* for on-wafer measurements. Generally, the reference plane is defined by the calibration procedure, with the reference plane occurring at the point at which the on-wafer standards are known. Ideally, this should be in a stretch of uniform transmission line, far enough from any transition or discontinuity (such as the pad-probe tip contact) that any evanescent waves have died away.

If the on-wafer transmission line has been characterized, the location of the reference plane can be translated through a section of uniform transmission line, provided that both the original and the translated planes are free of significant evanescent modes. If evanescent modes may be present, as in the case of a probe-tip calibration, it is necessary to employ a model to account for their effects. In that case (or if their effects are just ignored), the accuracy of any results will depend on the accuracy of the model (or the assumption that the effects can be ignored). Thus in estimating the uncertainty, one must evaluate the uncertainty introduced by the model and its underlying assumptions. We shall revisit this issue when we discuss on-wafer transistor measurements below.

5.3 Noise-Temperature Measurements

Although noise-temperature measurements are not the principal on-wafer noise measurements, they are of potential interest, and they provide a foundation for the more common noise-parameter measurements that are treated in Section 5.4. Figure 5.1 depicts the configuration for an on-wafer noise temperature measurement, with relevant reference planes indicated.

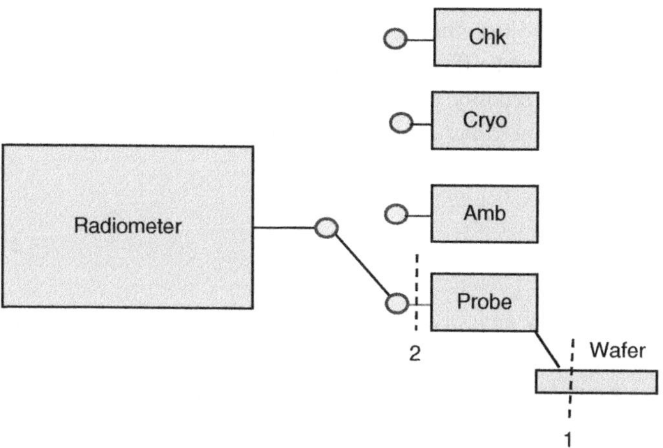

Figure 5.1 Configuration for on-wafer noise-temperature measurement.

The DUT is located on the wafer, and its noise temperature is to be determined at the on-wafer reference plane 1. The noise temperature at plane 2, T_2, can be measured by the radiometer using methods discussed in Chapter 3. The noise temperature at plane 1, T_1, is then computed from Eq. (1.28),

$$T_2 = \alpha_{21} T_1 + (1 - \alpha_{21}) T_{amb} \tag{5.1a}$$

$$T_1 = \frac{T_2 - (1 - \alpha_{21}) T_{amb}}{\alpha_{21}} \tag{5.1b}$$

where the probe is assumed to be at ambient temperature. The available power ratio α_{21} is given by

$$\alpha_{21} = \frac{|S_{21}|^2 \left(1 - |\Gamma_1|^2\right)}{|1 - \Gamma_1 S_{11}|^2 \left(1 - |\Gamma_2|^2\right)} \tag{5.2}$$

where Γ_1 is the reflection coefficient of the on-wafer DUT at plane 1, Γ_2 is the reflection coefficient from the probe at plane 2, and the S-parameters are those of the probe from plane 1 to plane 2. The location of the on-wafer plane 1 is determined by the on-wafer calibration, which will be discussed in more detail in Section 5.4.1.

The uncertainty analysis for on-wafer noise-temperature measurements is as outlined in Section 3.4,

$$u(T_1) \approx \sqrt{\left(\frac{u(T_2)}{\alpha_{21}}\right)^2 + (T_1 - T_{amb})^2 \left(\frac{u(\alpha_{21})}{\alpha_{21}}\right)^2} \tag{5.3}$$

The uncertainty in T_2 is just the uncertainty in a "normal" off-wafer noise-temperature measurement. If everything is reasonably well-matched, the contributions to the uncertainty due to the uncertainties in Γ_2, Γ_1, and S_{11} will be negligible. In that case, $u(\alpha_{21})$ is dominated by the uncertainty in S_{21},

$$\frac{u(\alpha_{21})}{\alpha_{21}} \approx \frac{u\left(|S_{21}|^2\right)}{|S_{21}|^2} \tag{5.4}$$

For measurements made at National Institute of Standards and Technology (NIST) [2], the resulting standard (1-σ) uncertainty in the measured on-wafer noise temperature at 8 GHz is plotted in Figure 5.2 as a function of the on-wafer noise temperature. Figure 5.2a shows the standard uncertainty, and Figure 5.2b shows the fractional uncertainty.

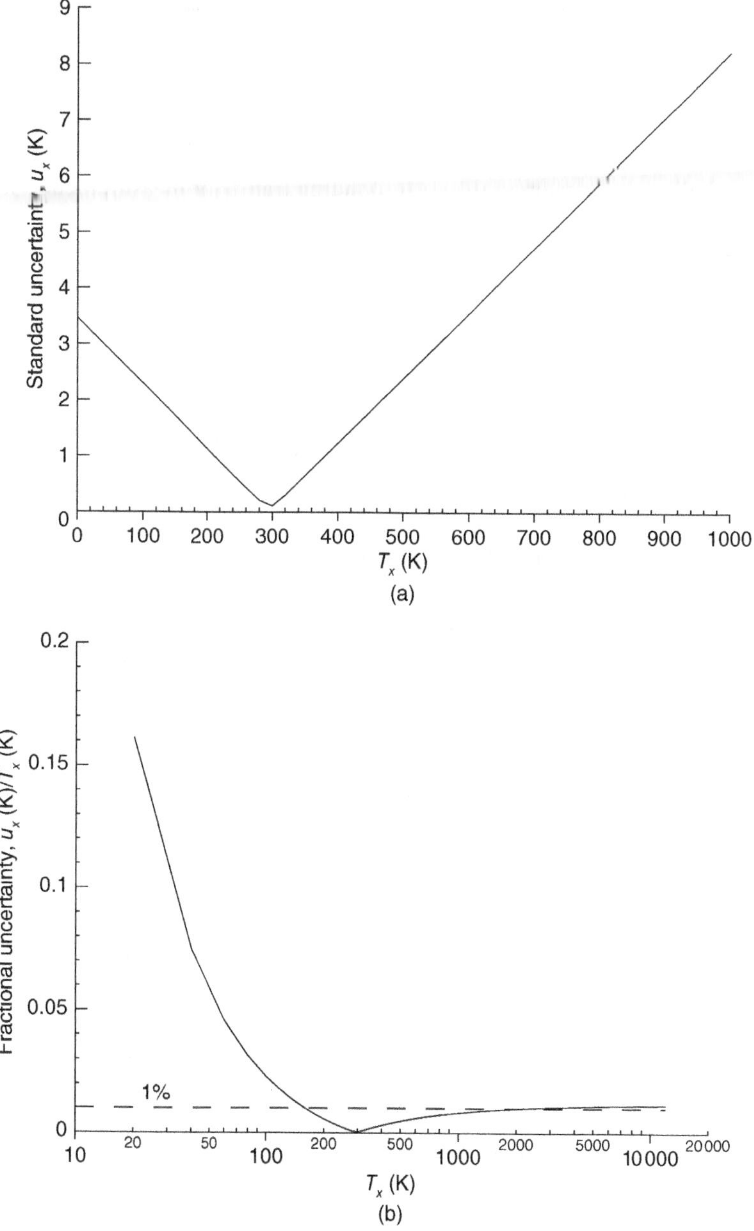

Figure 5.2 (a) Standard uncertainty in on-wafer noise-temperature measurements [2]. (b) Fractional uncertainty in on-wafer noise-temperature measurements. Source: [2]/U.S. Department of Commerce/Public Domian.

5.4 On-Wafer Noise-Parameter Measurements

5.4.1 General

The principal on-wafer noise measurements are those of transistor or amplifier noise parameters. On-wafer measurement of the noise parameters of amplifiers is just a less challenging version of the corresponding transistor measurements, so we will restrict our discussion to transistor measurements.

In principle, transistor noise-parameter measurements are much like noise-parameter measurements on packaged amplifiers, and similar methods should apply. In treating a transistor as a two-port device, the input (1) and output (2) ports are defined as in Figure 5.3. There are, however, several differences between measurements on transistors and on packaged amplifiers, which result in significant complications. Transistors generally have several properties that make them more problematic than packaged amplifiers. They are usually more susceptible to damage from stray static charges or from incorrect biasing, and thus they require considerable care in their handling and usage. Also, whereas amplifiers are usually designed to be well matched to a 50 Ω environment, transistors often have rather large values of S_{11} and S_{22}. In addition, transistors generally have lower noise levels than do packaged amplifiers, and consequently they are more likely to strain the resolution of a measurement system or method. Besides these properties of the DUTs themselves, the measurement environment introduces further complications. Transistors are typically measured on a wafer, requiring an on-wafer VNA calibration and removal of any probe and pad effects. There are potential problems associated with obtaining good contact between the probe tips and the contact pads on the wafer, and also due to possible vibrations of the probe station. Furthermore, it is often desired to know the transistor noise parameters referred to a reference plane as close as possible to the transistor terminals, which requires "de-embedding" the transistor from any transmission lines or other structure between it and the reference plane of the on-wafer calibration. Finally, the on-wafer environment restricts the achievable input reflection coefficients if they are generated by off-wafer terminations and transferred through the probe. Losses in the probe will prevent access to the outer edge of the unit circle, near $|\Gamma_1| = 1$.

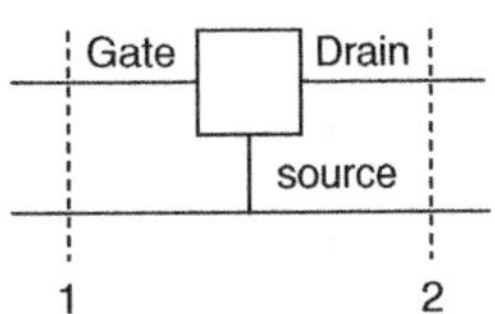

Figure 5.3 Definition of reference planes for a transistor.

Despite all these difficulties, good measurements of transistor noise parameters can be made. A general measurement setup is shown in Figure 5.4, with relevant reference planes indicated. The setup is similar to that for amplifiers in Figure 4.7, except that the DUT must be accessed through the two probes. As in Figure 4.7, the input terminations can be discrete terminations, or they can be achieved by using a tuner. The VNA is used to measure reflection coefficients of the input states

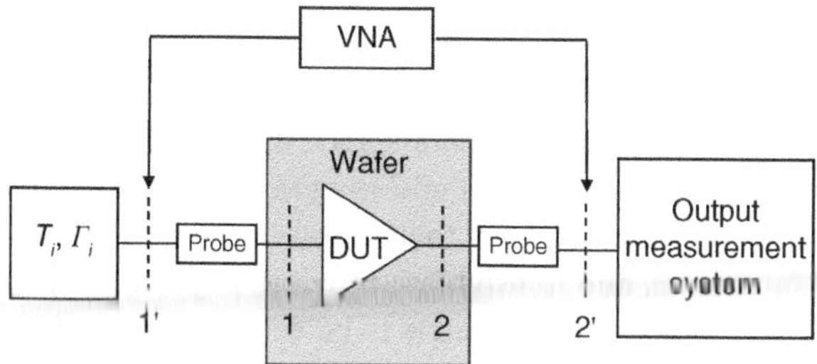

Figure 5.4 General setup for measuring noise parameters of an on-wafer transistor.

at plane 1, $\Gamma_{1,i}$, the output reflection coefficients at plane 2, $\Gamma_{2,i}$, and the DUT S-parameters, as well as the reflection coefficients at planes 1′ and 2′ if these are needed in the processing. (We will discuss the need – or lack thereof – for the 1′ and 2′ reflection coefficients further in Sections 5.4.2 and 5.5.1.) In order to measure reflection coefficients and S-parameters at planes 1 and 2 on the wafer, a two-tier [5, 6] calibration is performed on the VNA.

The location of the on-wafer reference planes warrants some discussion. The two-tier calibration just mentioned employs an on-wafer set of standards including a "through" standard. Ideally, the on-wafer standards are fabricated on the same wafer as the DUT, which is the case for the sample measurements presented later in this section. The reference plane for the calibration is taken to be the center of the "through" standard. The calibration is then translated back to the plane D, which is displaced by a distance (12 μm in the case shown) corresponding to the gap left for the transistor, as indicated in Figure 5.5. The translation is done using the characteristics of the transmission line, which are also determined in the calibration. (Only the reference planes on the left side are shown in Figure 5.5, but there are corresponding planes on the right side as well.) Extraction of the parameters at reference plane T in the figure would require additional standards at T. Measurement results presented in this section are at plane D.

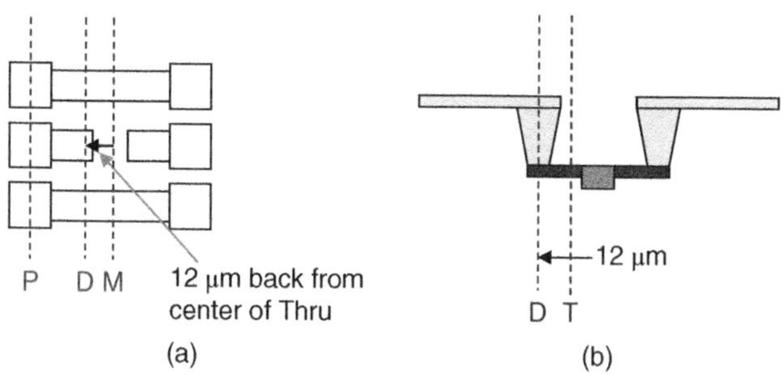

Figure 5.5 On-wafer reference planes: (a) from above and (b) from the side. Source: From Randa et al. [7]/IEEE.

5.4.2 Radiometer-Based Systems

There are good theoretical models for many types of transistors (e.g. see [8–11]). Consequently, when measuring transistor noise parameters, it is quite common to work within the context of a particular model and to perform the measurements necessary to determine the free parameters that appear in the model. This usually simplifies the measurements, and it also often provides physical insight into the physics of the transistor, since the parameters to be determined are usually physical parameters of the device. Nonetheless, there is still a need for full, model-independent measurements of transistor noise parameters, in order to verify models, to extend models into new parameter and frequency ranges, and to provide a foundation on which to build models of new transistor designs. In this section, we will review full, model-independent measurements, in which the DUT is defined as a particular device with fixed bias conditions. In Section 5.4.4, we will discuss enhanced, model-assisted methods.

The noise-parameter measurement procedure for a radiometer-based system, such as that depicted in Figure 5.6, is similar to the amplifier measurements discussed previously, with minor differences to account for the probes. Two standards (Std 1 and Std 2) are used to calibrate the radiometer, with an (optional) check standard (Chk) to confirm the calibration. A series of terminations $T_{i'}$, $\Gamma_{i'}$ are connected to the input probe, producing previously measured or computed noise temperatures and reflection coefficients $T_{1,i}$, $\Gamma_{1,i}$, at the transistor input, reference plane 1. The output noise temperature at plane 2′, $T_{2',i}^{meas}$, is measured for each input termination, and the output noise temperature at the on-wafer plane 2 is computed from

$$T_{2,i}^{meas} = \frac{T_{2',i}^{meas} - \left(1 - \alpha_{2'2}\right) T_{amb}}{\alpha_{2'2}} \tag{5.5}$$

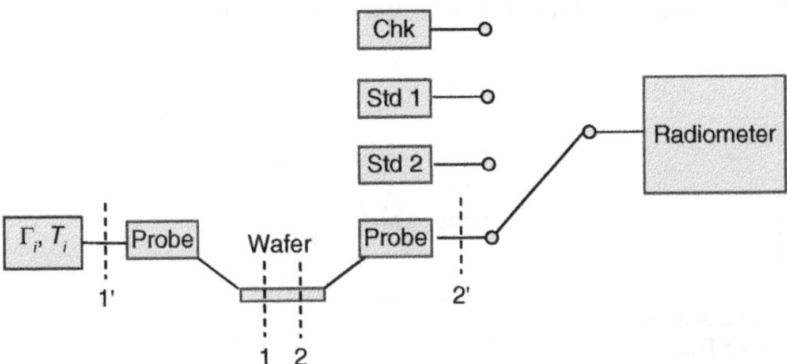

Figure 5.6 Outline of radiometer-based system for on-wafer noise-parameter measurements.

where $\alpha_{2'2}$, the available power ratio from plane 2 to plane 2′, is given by

$$\alpha_{2'2} = \frac{|S_{21}(2'2)|^2\left(1 - |\Gamma_{2,i}|^2\right)}{|1 - \Gamma_{2,i}S_{11}(2'2)|^2\left(1 - \left|\Gamma_{2',i}\right|^2\right)} \tag{5.6}$$

where $S_{ij}(2'2)$ is the S-matrix from plane 2 to plane 2′. These forward measurements can be supplemented by a measurement of the reverse noise temperature, as depicted in Figure 4.6. Such an additional measurement requires altering the measurement setup (turning around the DUT), and so it is seldom used for routine characterization measurements, but it can be useful for problem cases.

Armed with the set of measured output noise temperatures, we determine the noise parameters as we did for amplifiers, by minimizing the χ^2 of Eq. (4.35). The functional form for $T_2(T_i, \Gamma_i)$ depends on the set of noise parameters that one chooses. In Section 4.5.3 we used a form of the usual IEEE noise parameters, so here we give the form for the wave representation,

$$\begin{aligned} T_{2,i}(T_i, \Gamma_i) = {} & \frac{|S_{21}|^2}{\left(1 - |\Gamma_{2,i}|^2\right)}\left\{\frac{\left(1 - |\Gamma_{1,i}|^2\right)}{|1 - \Gamma_{1,i}S_{11}|^2}T_{1,i}\right. \\ & \left. + \left|\frac{\Gamma_{1,i}}{1 - \Gamma_{1,i}S_{11}}\right|^2 X_1 + X_2 + 2Re\left[\frac{\Gamma_{1,i}X_{12}}{1 - \Gamma_{1,i}S_{11}}\right]\right\} \end{aligned} \tag{5.7}$$

where the S-parameters are those of the DUT, from plane 1 to plane 2. The minimization process will determine the values for X_1, X_2, X_{12}, as well as $|S_{21}|^2$ (if desired). It is worth noting that this process requires measurements at planes 1 and 2 on-wafer ($\Gamma_{1,i}$, $\Gamma_{2,i}$ $T_{1,i}$, S_{ij}) and plane 2′ ($\Gamma_{2',i}$, $S_{ij}(2'2)$, $T_{2',i}$). If the usual IEEE parameters are desired, they can be computed using Eqs. (4.22) and (4.23). As with the packaged-amplifier case, the analysis and measurements can be done in terms of delivered power, rather than noise temperature (available power), by following the discussion at the end of Section 4.5.3.

Some sample results of on-wafer measurements of transistor noise parameters are shown in Figure 5.7, taken from [12]. The results shown are for the noise parameters in the wave representation; the corresponding IEEE parameters will be shown in Section 5.4.3. Because all four noise parameters in the wave representation have dimensions of temperature, they are all plotted on a single graph. The device measured was an NMOS transistor with 0.12 μm gate length. It possessed several features that make transistor measurements challenging: it was poorly matched, had very low noise temperature, and a very large value of $|\Gamma_{opt}|$. This will be discussed further in Section 5.4.3.

A total of nine forward states and one reverse measurement were used to obtain these results. The forward states included one well-matched hot source (around 1100 K) and eight ambient temperature terminations chosen to produce adequate coverage of the complex plane over the entire frequency range. The complex input

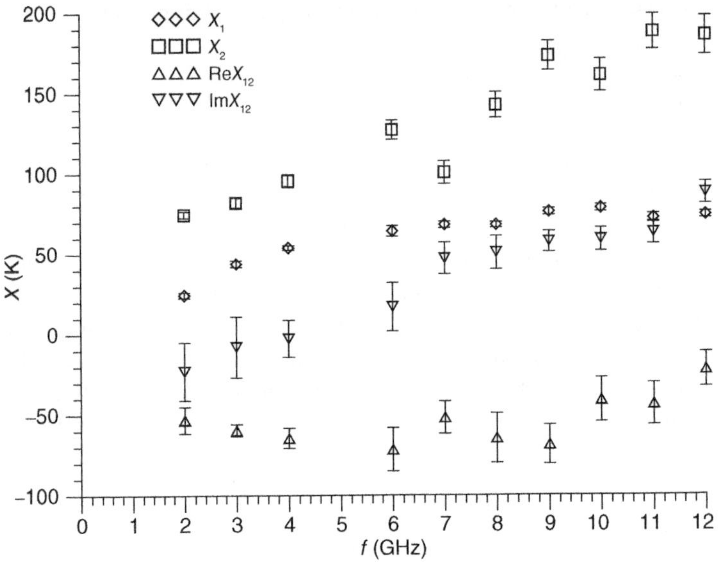

Figure 5.7 Wave-representation results for noise parameters of a particular transistor. Source: From Randa and Walker [12]/IEEE.

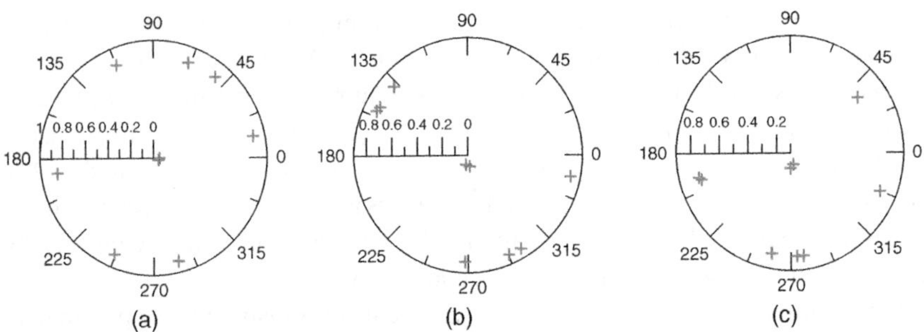

Figure 5.8 Reflection coefficients of input terminations at (a) 2 GHz, (b) 7 GHz, and (c) 12 GHz. Source: From Randa [13]/John Wiley & Sons.

reflection coefficients are shown in Figure 5.8 for three representative frequencies. Because the reflection coefficients vary with frequency, using a fixed set of terminations leads to a better distribution of the input reflection coefficients at some discrete frequencies than at others.

The uncertainties in Figure 5.7 are the combined standard uncertainties [14], including both type A (statistical) and type B (all other) contributions. The uncertainties on the quantity X_1 are particularly small because it is determined very well by the reverse measurement. Note that X_2 is the effective input noise temperature for the matched case ($\Gamma_{1,i} = 0$). It varies from about 70 K to around 180 K.

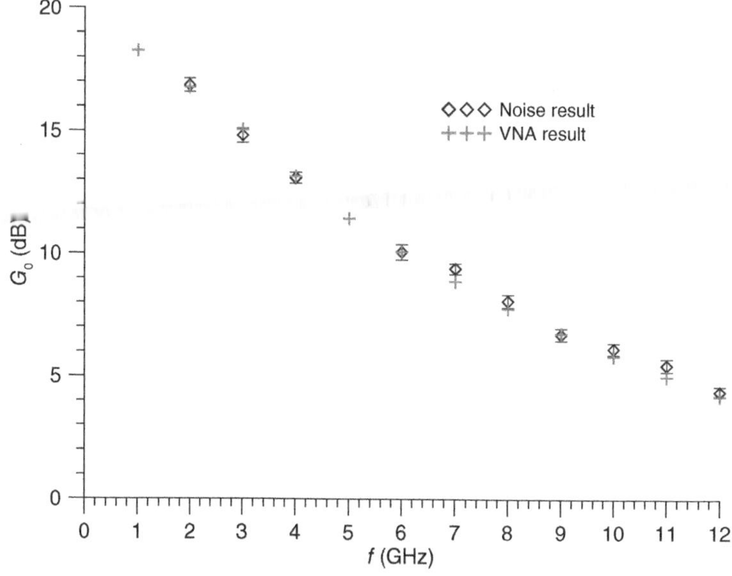

Figure 5.9 Comparison of noise and VNA determinations of $G_0 = |S_{21}|^2$ for a particular transistor. Source: From Randa and Walker [12]/IEEE.

The quantity $G_0 \equiv |S_{21}|^2$ between planes 1 and 2 was also determined in the fitting process. Since G_0 was measured with the VNA as well, we can compare the two sets of results as a check. This is done in Figure 5.9 for G_0 in decibels. The good agreement between the two independent sets of measurements constitutes a partial check of the results. It is interesting to note that the small disagreement in the G_0 results at 7 GHz coincides with a suspicious low point in the X_2 results.

5.4.3 Commercial Systems and Reference-Plane Considerations

Commercial systems for on-wafer transistor noise-parameter measurements are similar to those for connectorized noise-parameter measurements, except, of course, that the reference planes are on a wafer. Typical measurement setups are sketched in Figure 5.10, with Figure 5.10b showing the on-wafer measurement calibration configuration, and Figure 5.10c showing the actual DUT-measurement configuration. In addition (and usually prior) to the on-wafer calibration, an initial system calibration is required, Figure 5.10a, to calibrate the combination of receiver and power meter. This initial system calibration also enables one to compute the reflection coefficients necessary to compute the mismatch factor at the DUT output reference plane.

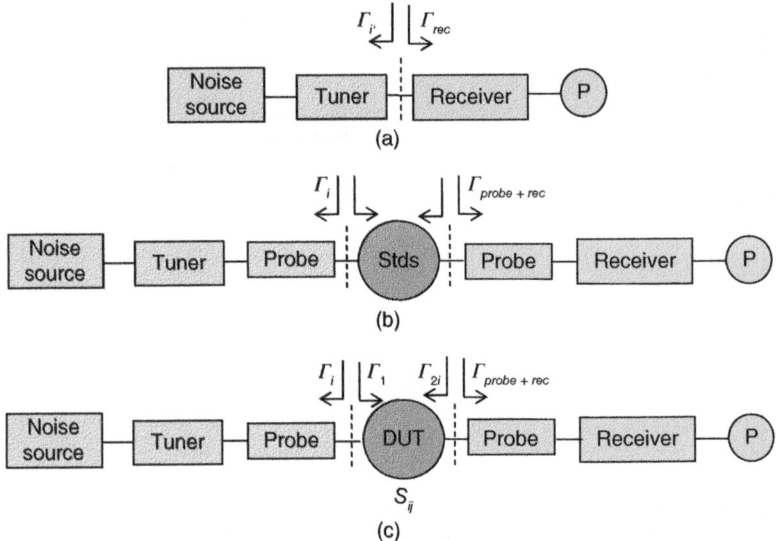

Figure 5.10 (a) System calibration. (b) On-wafer calibration configuration. (c) DUT measurement configuration.

In commercial systems, the on-wafer calibration is typically done at the probe tip, indicated by P (or P_1 and P_2) in Figures 5.5 and 5.11. For convenience, the probe-tip calibration often uses a set of standards on a separate wafer than the DUT. In order to move the reference planes closer to the transistor, it is necessary to either measure or model the characteristics of the structure between the probe tips and the planes of interest. The two-tier calibration discussed above allows translation to the D reference planes. It is also possible to fabricate and measure structures with an open or a short between planes T_1 and T_2 in Figure 5.11. These measurements, supplemented by model calculations can be used to "deembed" the transistor properties (between T_1 and T_2) from the probe-tip measurement data [15].

It is instructive to compare measurement results obtained with commercial systems to those obtained with the radiometer-based system of in Section 5.4.2. This is done in Figure 5.12a–d for the IEEE noise parameters [7]. The Lab-A and Lab-B results were obtained with commercial systems, whereas the NIST measurements were performed on a radiometer-based system. The frequency ranges of the measurements at the three labs were different due to the limitations of the different systems. All

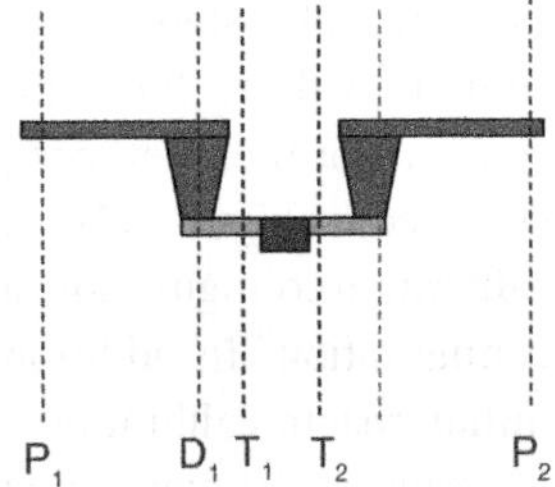

Figure 5.11 On-wafer reference planes.

three sets of measurements used a set of calibration standards fabricated on the same wafer as the DUT, and all the results in Figure 5.12 are at the reference plane D of Figure 5.5. The error bars on the NIST results correspond to the standard uncertainty (one sigma). Uncertainty estimates were not available for the Lab-A and Lab-B results, but it would be expected that they would be comparable to, or perhaps somewhat larger than, the NIST uncertainties. The DUT was the same device as of Section 3.4.2 and thus the NIST results correspond to the X parameters of Figure 5.7. This device posed significant measurement challenges. It had a value of $|S_{11}|$ above 0.9 at 2 GHz and above 0.5 throughout the measurement range. It was very low noise, with a value for T_e below 100 K throughout the range, and indistinguishable from zero at the lowest frequencies. Also, its $|\Gamma_{opt}|$ was near one at the lower frequencies and above 0.5 throughout the frequency range.

For the minimum noise figure in dB (F_{min}), graphed in Figure 5.12a, there is a considerable spread among the measured values, with differences of around 0.5 dB at the low frequencies. The NIST uncertainties and the differences among the measurements at the different laboratories indicate that F_{min} is smaller than the collective resolution at low frequencies. For R_n in Figure 5.12b, there is good agreement at the higher frequencies, but there appears to be a systematic divergence of one or two ohms at the lowest frequencies. The situation for $|\Gamma_{opt}|$, Figure 5.12c, is rather muddled at the lowest frequencies, where it is very difficult to measure due to its large value (0.9 or above). At the higher frequencies, the NIST and Lab-A results differ by about 0.1 or a little more. Figure 5.12d shows that the measurements of ϕ_{opt}, the phase of Γ_{opt}, at all three laboratories are in very good agreement, except for one point.

Given what we know about the uncertainties, the agreement among the different labs and different measurement systems is not so bad, but neither is it good. If the commercial-system results are assigned uncertainties comparable to those of the NIST measurements, then for the most part, the results at the different laboratories are in agreement. However, it is evident that the uncertainties need to be (or at least needed to be at the time of those measurements) considerably smaller to measure these devices at frequencies below about 6 GHz. Without some improvement in the measurement methods, the noise performance of very low-noise transistors will continue to be better than the resolution of our (model-independent) measurements of them.

5.4.4 "Enhanced" or Model-Assisted Measurements

As noted above, there are good models for common transistor types, and measurements are often performed in the context of such models in order to simplify or extend the measurements and to directly determine underlying physical

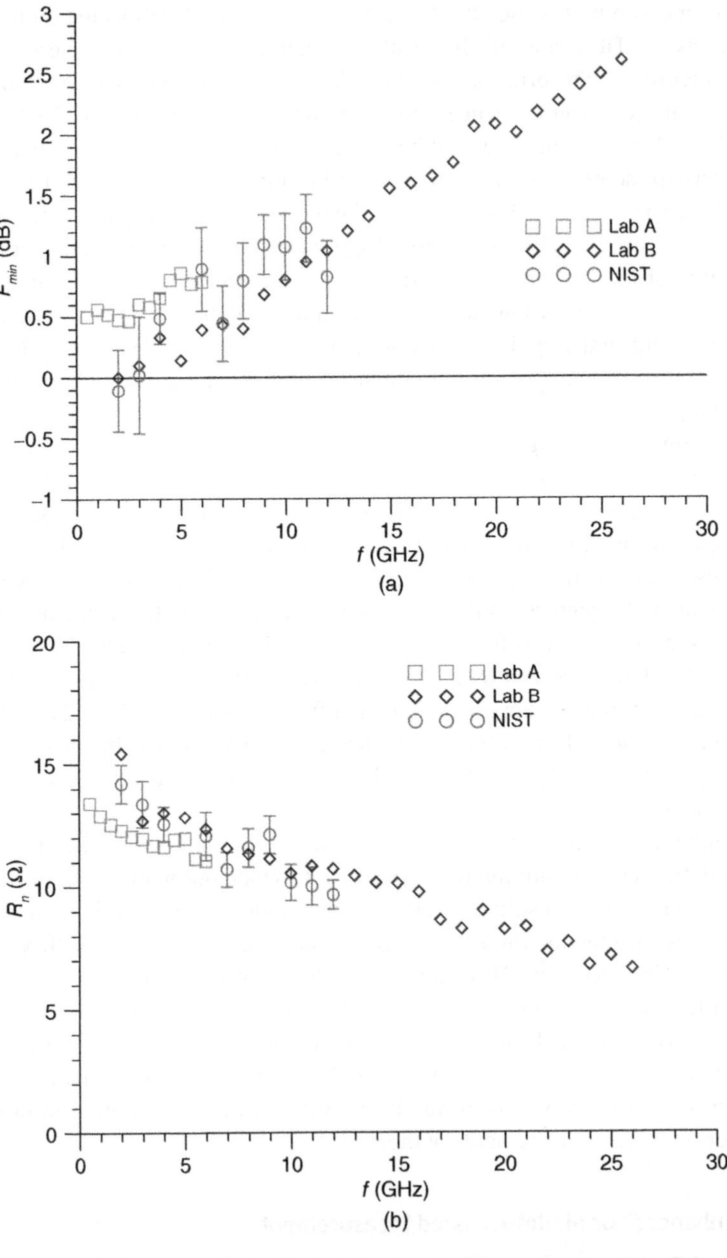

Figure 5.12 (a) Comparison of measurement results at reference plane D for F_{min} of an on-wafer transistor. (b) Comparison of measurement results for R_n of an on-wafer transistor. (c) Comparison of measurement results for $|\Gamma_{opt}|$ of an on-wafer transistor. (d) Comparison of measurement results for the phase of Γ_{opt} of an on-wafer transistor. Source: From Randa et al. [7]/IEEE.

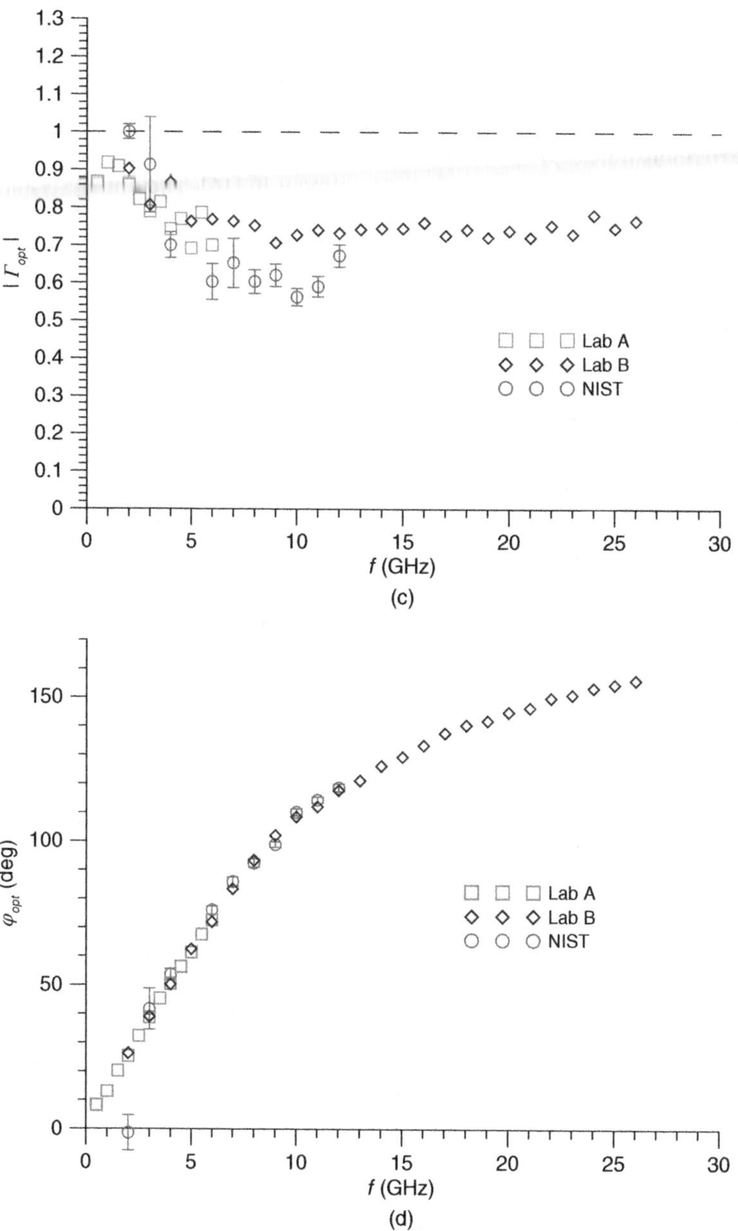

Figure 5.12 (*Continued*)

parameters of the transistor. Also, it is possible to exploit features of the frequency dependence or time-domain properties of the measurements in order to simplify the measurements. There are numerous models and methods that have been used in this manner, and it is beyond the scope of this book to review them all. We will instead discuss measurements within the context of one particular model in order to demonstrate some of the ideas.

An early example of this approach is the treatment in [16]. Their analysis and modeling proceeds in two steps. First the modeling and analysis of Van Der Ziel [17, 18] is used to show that two noise parameters (R_n and $|Y_{opt}|$) can be determined by measuring the noise figure for a matched input termination as a function of frequency. Then a more detailed model for field effect transistor (FET) noise [19, 20] is used to compute the remaining noise parameters without the need for additional measurements.

The first step begins with the representation of a noisy two-port as a voltage noise source $\langle |v_n|^2 \rangle$ and a current noise source $\langle |i_n|^2 \rangle$ at the input of a noiseless two-port, as in Figure 4.3. The entire noisy two-port is then represented by Figure 5.13a, which separates the source and gate resistances (R_s and R_g) from the intrinsic (noisy) transistor. (It is assumed that the measurement plane is on-wafer and that one has calibrated out or accounted for any effect of the contact pads.) The noisy intrinsic transistor is then represented by an intrinsic noiseless transistor with gate and drain current noise sources, as in Figure 5.13b. The parameters of Figure 5.13a

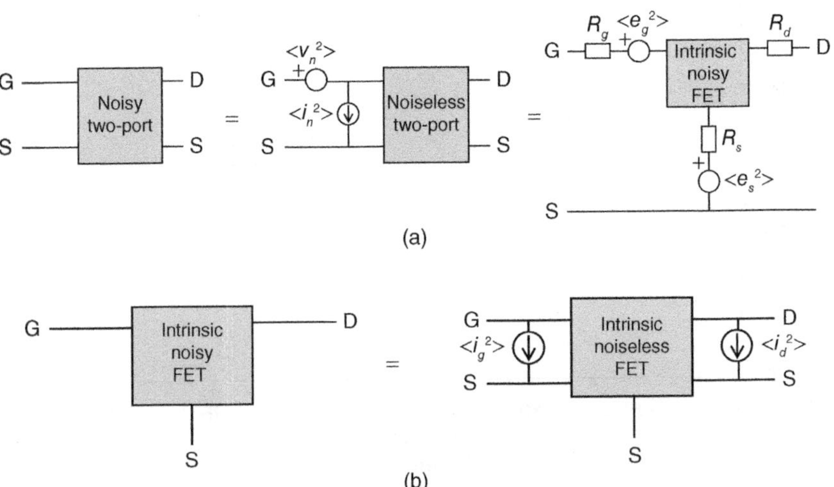

Figure 5.13 (a) Transistor with parasitic resistances extracted. (b) Thevenin equivalent circuit for a noisy FET.

are related to those of Figure 5.13b by

$$\left\langle |v_n|^2 \right\rangle = \frac{\left\langle |i_d|^2 \right\rangle}{|Y_{21}|^2} + 4k_B T_0 (R_s + R_g)\Delta f$$

$$\left\langle |i_n|^2 \right\rangle = \left\langle |i_g|^2 \right\rangle + \frac{|Y_{11}|^2 \left\langle |i_d|^2 \right\rangle}{|Y_{21}|^2} - 2Re\left(\frac{Y_{11} \left\langle i_g^* i_d \right\rangle}{Y_{21}} \right)$$

$$\left\langle v_n^* i_n \right\rangle = \frac{Y_{11} \left\langle |i_d|^2 \right\rangle}{|Y_{21}|^2} - \frac{\left\langle i_g i_d^* \right\rangle}{Y_{21}^*} \tag{5.8}$$

Using the Van Der Ziel results, one can then show [16] that R_n is frequency independent, and that B_{cor} is proportional to frequency and G_{cor} is proportional to frequency squared. We then use

$$F(Y_S) = F_{min} + \frac{R_n}{G_S}|Y_S - Y_{opt}|^2$$

$$F_{min} = 1 + 2R_n(G_{opt} + G_{cor}) \tag{5.9}$$

and consider the matched case, $F_0 \equiv F(Y_S{=}Y_0)$,

$$F_0 = 1 + R_n G_0 + \frac{R_n}{G_0}\left[2G_0 G_{cor} + |Y_{opt}|^2\right] \tag{5.10}$$

As a function of frequency, R_n and G_0 are constant, whereas G_{cor} and $|Y_{opt}|^2$ are proportional to frequency squared. Therefore, if we measure F_0 as a function of frequency, the extrapolated intercept $F_0(\omega = 0)$ determines R_n,

$$F_0(\omega = 0) = 1 + R_n G_0 \tag{5.11}$$

Furthermore, the value of $|Y_{opt}|$ can be determined from the slope of F_0 versus ω^2. Thus, the measurement of the noise figure as a function of frequency suffices to determine two of the four noise parameters, R_n and $|Y_{opt}|$, or equivalently, $\langle |v_n|^2 \rangle$ and $\langle |i_n|^2 \rangle$.

In order to determine the remaining two noise parameters, one can revert to the direct measurement methods, as described in Sections 5.4.2 and 5.4.3, or use additional model information. If direct measurements are used, the fact that there are only two, rather than four, remaining noise parameters to be determined means that fewer measurements should be needed. If additional model information is used, no additional measurements are needed. In [16], the additional information was based on the model proposed in [19] and [20], represented in Figure 5.14. In this model, the noise properties of the FET are determined by two uncorrelated noise sources, represented by R_{gs} at temperature T_g and G_{ds} at temperature T_d.

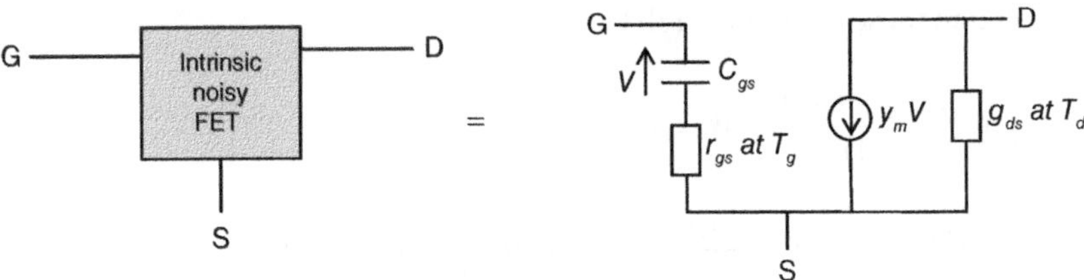

Figure 5.14 Simple model for a noisy intrinsic FET.

In the context of this model, it turns out that the correlation between the gate and drain noise currents (Van der Ziel representation) is given by

$$\frac{\langle i_g i_d^* \rangle}{\sqrt{|i_g|^2 |i_d|^2}} \approx j \frac{|Y_{21}|}{|Y_{11}|} \sqrt{\frac{\langle |i_g|^2 \rangle}{\langle |i_d|^2 \rangle}} \tag{5.12}$$

Along with our knowledge of R_n and $|Y_{opt}|$ (or equivalently, $\langle |v_n|^2 \rangle$ and $\langle |i_n|^2 \rangle$), Eq. (5.12) allows us to determine the remaining noise parameters as follows. We can deduce $\langle |i_d|^2 \rangle$ from $\langle |v_n|^2 \rangle$ by using Eq. (5.8). Using the value of $\langle |i_d|^2 \rangle$ and Eq. (5.12) in the expression for $\langle |i_n|^2 \rangle$ in Eq. (5.8) enables us to determine $\langle |i_g|^2 \rangle$, and similarly Eq. (5.12) enables us to determine $\langle v_n^* i_n \rangle$ in Eq. (5.8).

Thus, within the context of this model, the problem of measuring the noise parameters has been reduced to the measurement of the matched noise figure F_0 as a function of frequency. In addition to the measurement simplification, considerable physical insight is gained by the determination of $\langle |i_g|^2 \rangle$ and $\langle |i_d|^2 \rangle$ and their relationships to the traditional noise parameters.

The obvious limitation of this method is that it is only as good as the model for the FET. And in addition, there are potential errors associated with measurement of F_0 with imperfectly matched terminations. Nonetheless, the considerable advantages of such an approach have resulted in its widespread use. This same basic model, with minor modifications, is still in use today.

Besides using physical or circuit models of transistor noise behavior, other approaches incorporate additional information or assumptions in other manners. As examples, a time-domain analysis of the Fourier-transformed noise power is applied in [21]; a smooth fit to the frequency dependence of the intrinsic noise matrix elements is used in [22]; and a corrected Y-factor method is developed and applied in [23]. Extraction of noise and scattering parameters from noise-figure measurements only and from VNA and noise-figure measurements in the context of a HEMT (high electron mobility transistor) model is presented in [24].

5.5 Uncertainties

5.5.1 Differences from Packaged Amplifiers

As in the off-wafer amplifier case in Chapter 4, the on-wafer noise-parameter uncertainties are best evaluated by a Monte Carlo computation [25]. Details can be found in [26], and some results are reflected in the sample measurement results shown earlier in this chapter. Although the analysis is similar to the packaged-amplifier case, there are a few additional complications. The measurement planes of interest are on the wafer, and the uncertainties in VNA and noise measurements are different on a wafer from what they are in coaxial lines. Thus, the input uncertainties are different for the on-wafer case. Also, for the output noise power or noise temperature, we must characterize and correct for the effects of the probes, which introduces additional uncertainty. This additional uncertainty is complicated by the fact that the transistor may be very poorly matched, leading to relatively large values of the reflection coefficient at its output. This requires that we refine our estimate of the uncertainty in measuring the output noise power or noise temperature. It also introduces a new complication: the fit to the noise measurements may sometimes result in unphysical values for the noise parameters. That requires us to adopt a prescription for handling unphysical results in the simulations. We will discuss each of these complications in turn.

The uncertainty in the noise temperatures of the ambient-temperature terminations is the same as in the coaxial case in Chapter 4. Reflection coefficients and S-parameters measured at on-wafer reference planes have larger uncertainties than those measured at reference planes in coaxial lines. The uncertainties to be used for the on-wafer reflection coefficients and S-parameters will depend on the specific methods and instrumentation. The input uncertainty used in [25] was $u(Re\Gamma) = u(Im\Gamma) = 0.005$ with a correlation coefficient of $\rho = 0.36$ in the 1–12 GHz range. A concern in any on-wafer measurement is the repeatability of the contact between the probe tip and the on-wafer contact pad. This enters the uncertainty analysis the same way as repeatability of coaxial connections in off-wafer measurements. It is included in the input S-parameter and reflection coefficient uncertainties.

The measurement planes of interest are on the wafer, as indicated in Figures 5.4 or 5.6, whereas the noise-temperature or power measurements are performed at plane 2′. One must therefore correct for the effect of Probe 2 to obtain the noise temperatures at plane 2. The S-parameters of the probe are determined in the two-tier calibration that is used to calibrate the VNA down to the on-wafer reference planes. To obtain the noise temperature at plane 2 ($T_{2,i}$) from the noise temperature at plane 2′ ($T_{2',i}$) we treat the probe as an adapter, characterized by its available-power ratio $\alpha_{2'2,i}$,

$$\alpha_{2'2,i} = \frac{|S_{2'2}|^2 \left(1 - |\Gamma_{2,i}|^2\right)}{|1 - \Gamma_{2,i} S_{22}|^2 \left(1 - |\Gamma_{2',i}|^2\right)} \tag{5.13}$$

where $S_{2'2}$ is the S-parameter of probe 2 from plane 2′ to plane 2, S_{22} is the reflection S-parameter of probe 2 at plane 2 (what would normally be called S_{11}), $\Gamma_{2,i}$ is the reflection coefficient at plane 2 (from the transistor), and $\Gamma_{2',i}$ is the reflection coefficient at plane 2′ (from the probe). Knowing $\alpha_{2'2,i}$, one can compute the output noise temperature at plane 2 in terms of $T_{2',i}$ in the usual manner,

$$T_{2,i} = \frac{T_{2',i} - \left(1 - \alpha_{2'2,i}\right) T_a}{\alpha_{2'2,i}} \tag{5.14}$$

where T_a is the noise temperature of the probe, which is assumed to be at ambient temperature.

From Eq. (5.14) we see that in order to compute the uncertainty in the on-wafer output noise temperature $T_{2,i}$, we need the uncertainties in T_a, $T_{2',i}$, and $\alpha_{2'2,i}$. The uncertainty in T_a is the same as in the case of the coaxial amplifier measurements, treated above. (As in that case the uncertainty in the ambient-temperature standard is different and is treated within the uncertainty in the measurement of the output noise temperature or power, as detailed in Chapter 3.) The probe effect, of course, was absent from the treatment of Chapter 4, and there is also a new complication with the noise measurement at the coaxial plane $T_{2',i}$. Because the on-wafer transistor may have a relatively large value of $|S_{22}|$, the reflection coefficient at the coaxial measurement plane $\Gamma_{2',i}$ can also be large. That may require special care in estimating the uncertainty in the power or noise-temperature measurement at plane 2′. Such detailed treatment is beyond the scope of the current exposition, but an example can be found in [26].

The final complication that arises in the uncertainty analysis for on-wafer noise parameter measurements is the occurrence of results that violate physical bounds. This is more common in on-wafer measurements than it is for packaged amplifiers because the uncertainties are greater on-wafer, and because the noise levels and reflection coefficients are often closer to the physical limits. In measurement simulations, as in actual measurements, it is possible that the fit to a set of measurements yields unphysical results, results which are obviously (or not so obviously) in error because they violate basic physical or mathematical constraints. Examples would include results with $T_{e,min} < 0$ or $|\Gamma_{opt}| > 1$. There are basically three ways of dealing with such cases in the Monte Carlo computation of uncertainties. (i) One could do a constrained fit, searching for the set of parameters that minimizes the fitting function and also satisfies all physical constraints. (ii) The simulated set of unphysical results could be included in the sample of simulated measurement sets. (iii) The unphysical measurement set could be discarded, on the basis that a set of real measurements that yielded unphysical results would be discarded and re-measured. There are arguments for and against each of these approaches [25, 26]. The results presented in Section 5.5.2 were obtained with approach 3.

There is another problem in the analysis of the simulated measurement sets that requires special treatment. It is possible that the set of simulated measurements (or a set of real measurements) does not admit a good fit, that the best fit yields a relatively large value of χ^2 per degree of freedom (ν). The obvious solution is to discard such a simulated measurement set, as one would for a real measurement set. For this purpose, a cut of $\chi^2/\nu \leq 1$ is reasonable.

5.5.2 General Features and Properties

The general features of which noise-parameter uncertainties are sensitive to which input uncertainties are similar to the packaged-amplifier case. As for actual values of the uncertainties, Figure 5.12a–d shows them for one particular case, where the error bars correspond to standard uncertainties. The approximate values for the uncertainties in those results for on-wafer noise parameters are: 0.22–0.30 dB for G_0, 0.2–0.3 dB for F_{min}, 15–25 K for T_{min}, 0.6–0.8 Ω for R_n, 0.02–0.07 for $|\Gamma_{\text{opt}}|$, and 1°–6° for ϕ_{opt}. As expected, the uncertainties in the on-wafer case are generally somewhat larger than for the packaged amplifier. As in the amplifier case above, if the measurements could be done without an output attenuator (those measurements used a 10 dB attenuator between the output probe and the radiometer), the uncertainties would be smaller.

Again, as in the amplifier case, the actual values for the uncertainties depend on any number of different variables, including the choice of input terminations, the uncertainties in the noise temperatures and reflection coefficients of those input terminations, uncertainties in the measurements of the output noise powers or noise temperatures, correlations among errors in all the measurements, uncertainties in the S-parameters of the DUT, and the properties of the DUT itself. Consequently, the uncertainties in Figure 5.12 should be considered as no more than anecdotal evidence of what might occur in a general case.

More representative results can be obtained by considering a range of different DUT properties, input uncertainties, and measurement strategies. This was done in [27], which presented the results of a detailed study of the uncertainties in on-wafer measurements of transistor noise parameters. Five different DUTs were used in the study, covering a range of values of the noise parameters. This was done to permit the extraction of general features, rather than obtain results that depended on some coincidence of the particular values used. The simulations were all performed with a rather idealized set of input terminations, appropriately truncated to remove any terminations for which the DUT was unstable. The distribution of input reflection coefficients was not one of the variables considered in the study. It was assumed that the gain parameter (represented by $G_0 \equiv |S_{21}|^2$) was determined by the noise measurements, rather than from VNA measurements of the S-parameters. The reference set of uncertainties for the input parameters was that used in actual noise-parameter measurements at NIST. Although the specific numerical values for the noise-parameter uncertainties will depend on the specific input values used and on the method for determining G_0, one would expect the qualitative features that are extracted from the results to apply more generally than the particular cases studied.

Reference [27] first investigated the dependence of the noise parameters on the input uncertainties, i.e. the uncertainties in the measurements of the various reflection coefficients, noise temperatures, scattering parameters, and the like. The dependence of the noise-parameter uncertainties on the input uncertainties exhibited some differences from the case of connectorized-amplifier noise parameters, due in part to the buffering effect of the input and output probes and in part to the more challenging measurement problem posed by the properties of the transistors – the very low noise levels and the very poor matching to a 50 Ω environment. The most striking difference between the off-wafer and on-wafer cases was that the noise-parameter uncertainties in the on-wafer case are more sensitive to the reflection-coefficient uncertainties and less sensitive to other input uncertainties than in the off-wafer case. In particular, roughly one quarter to one third of the uncertainties in T_{min}, R_n, and G_0 are due to uncertainties in the input hot temperature and the probe loss. The uncertainties in measuring the output noise temperatures are responsible for roughly 5–15% of the uncertainties in T_{min}, R_n, and G_0. The remainders are due to uncertainties in the input reflection coefficients. For Γ_{opt} the fraction of the uncertainty due to reflection-coefficient errors is roughly 70–80% for all the DUTs considered. (All these fractions were for the set of input uncertainties that were used in the standard analysis; if one of the input uncertainties were increased or decreased, the corresponding fractions would of course change accordingly.)

5.5.3 Measurement Strategies

Reference [27] also used simulations to study the efficacy of various strategies for reducing the noise-parameter uncertainties in on-wafer measurements, focusing on those strategies that were found effective for connectorized amplifiers. The results for on-wafer transistors differed significantly from those for connectorized amplifiers. For connectorized amplifiers, adding an input termination with a cryogenic noise temperature led to a significant improvement in the uncertainties, particularly for T_{min} and G_0 but also for R_n. For on-wafer transistors, the effect of adding a cryogenic input termination is blunted by the presence of the input probe, and the improvements are considerably smaller, though still present. The addition of a reverse measurement, which helped considerably in the off-wafer case, is also quite effective in reducing the uncertainties on wafer, particularly for Γ_{opt}. As in the off-wafer case, however, it is somewhat inconvenient to implement. Other possible improvements were also studied. Using probes with less loss had minimal direct effect, while the availability and use of higher values for the maximum magnitude of the input reflection coefficient led to a small improvement in the uncertainty for Γ_{opt}. As was the case for connectorized-amplifier measurements, much smaller uncertainties resulted from measuring the reflection coefficient at the output of the DUT, rather than computing it from a cascade of input reflection coefficient and DUT S-parameters.

References

1 N. Markuvitz, *Waveguide Handbook*, Chapter 1, McGraw-Hill, New York, 1951; Peter Peregrinus, London, 1986.

2 J. Randa, "Noise temperature measurements on wafer," NIST Technical Note **1390** (March 1997). https://doi.org/10.6028/NIST.TN.1390 (accessed 9 July 2022).

3 R. Marks and D. Williams, "A general waveguide circuit theory," *J. Res. Natl. Inst. Stand. Technol.*, **97**,no. 5, pp. 533–562 (September–October 1992).

4 NIST Microwave Uncertainty Framework, Beta Version, 2011, [online] Available: https://www.nist.gov/services-resources/software/wafer-calibration-software (accessed 9 July 2022).

5 R. Marks, "A multi-line method of network analyzer calibration," *IEEE Trans. Microw. Theory Tech.*, **39**, no. 7, pp. 1205–1215 (July 1991).

6 D. Williams and R. Marks, "Transmission line capacitance measurement," *IEEE Microw. Guided Wave Lett.*, **1**, no. 9, pp. 243–245 (September 1991).

7 J. Randa, S. Sweeney, T. McKay, D.K. Walker, D. Greenberg, J. Tao, J. Mendez, G.A. Rezvani, and J. Pekarik, "Interlaboratory comparison of noise-parameter measurements on CMOS devices with 0.12 μm gate length," in *66th ARFTG Conference Digest*, pp. 77–81 (December 2005). doi: 10.1109/ARFTG.2005.8373127.

8 F. Danneville, "Microwave noise and FET devices," *IEEE Microwave Mag.*, **11**, no. 6, pp. 53–60 (October 2010).

9 M.W. Pospieszalski, "Interpreting transistor noise," *IEEE Microwave Mag.*, **11**, no. 6, pp. 61–69 (October 2010).

10 M.J. Deen, D.-H. Chen, S. Asgaran, G.A. Rezvani, J. Tao, and Y. Kiyota, "High-frequency noise of modern MOSFETs: compact modeling and measurement issues," *IEEE Trans. Electron Devices*, **53**, no. 9, pp. 206–2081 (September 2006).

11 C.-H. Chen, "Thermal noise measurement and characterization for modern semiconductor devices," *IEEE Instrumentation & Measurement Magazine*, **24**, no. 2, pp. 60–71 (April 2021). doi: 10.1109/MIM.2021.9400958.

12 J. Randa and D.K. Walker, "On-wafer measurement of transistor noise parameters at NIST," *IEEE Trans. Instrum. Meas.*, **56**, no. 2, pp. 551–554 (April 2007). doi: 10.1109/TIM.2007.891145.

13 J. Randa, "Amplifier and transistor noise-parameter measurements," in *Wiley Encyclopedia of Electrical and Electronics Engineering*ed. J. Webster (2014). doi: 10.1002/047134608X.W8219.

14 ISO *Guide to the Expression of Uncertainty in Measurement*, International Organization for Standardization, Geneva, Switzerland, 1993.

15 C.-H. Chen and M.J. Deen, "RF CMOS noise characterization and modeling," in *CMOS RF Modeling, Characterization, and Applications*, eds. M.J. Deen and T.A. Fjeldly, World Scientific, River Edge, NJ, 2002.

16 G. Dambrine, H. Happy, F. Danneville, and A. Cappy, "A new method for on wafer noise measurement," *IEEE Trans. Microw. Theory Tech.*, **41**, no. 3, pp. 375–381 (1993).

17 A. Van Der Ziel, "Thermal noise in field effect transistor," *Proc. IRE*, **50**, pp. 1808–1812 (1962).

18 A. Van Der Ziel, "Gate noise in field effect transistors at moderately high frequencies," *Proc. IRE*, **51**, pp. 461–467 (1963).

19 M.S. Gupta, O. Pitzalis, S.E. Rosembaum, and P. Greiling, "Microwave noise characterization of GaAs MESFET's by on-wafer measurement of output noise current," *IEEE Trans. Microw. Theory Tech.*, **35**, no. 12, pp. 1208–1217 (December 1987).

20 M.W. Pospieszalski, "Modeling of noise parameters of MESFETs and MODFFEs and their frequency and temperature dependence," *IEEE Trans. Microwave Theory Tech.*, **MTT-37**, pp. 1340–1350 (September 1989).

21 T. Werling, E. Bourdel, D. Pasquet, and A. Boudiaf, "Determination of wave noise sources using spectral parametric modeling," *IEEE Trans. Microwave Theory Tech.*, **45**, no. 12, pp. 2461–2467 (December 1997).

22 A. Lázaro, L. Pradell, and J.M. O'Callaghan, "FET noise-parameter determination using a novel technique based on 50-Ω noise-figure measurements," *IEEE Trans. Microwave Theory Tech.*, **47**, no. 3, pp. 315–324 (March 1999).

23 L.F. Tiemeijer, R.J. Havens, R. de Kort, and A.J. Scholten, "Improved *Y*-factor method for wide-band on-wafer noise-parameter measurements," *IEEE Trans. Microwave Theory Tech.*, **53**, no. 9, pp. 2917–2925 (September 2005).

24 A. Caddemi, A. Di Paola, and M. Sannino, "Full characterization of microwave low-noise packaged HEMT's: measurements versus modeling," *IEEE Trans. Instrum. Meas.*, **46**, no. 2, pp. 490–494 (April 1997).

25 J. Randa, "Uncertainty analysis for noise-parameter measurements at NIST," *IEEE Trans. Instrum. Meas.*, **58**, no. 4, pp. 1146–1151 (April 2009). doi: 10.1109/TIM.2008.2007044.

26 J. Randa, "Uncertainty analysis for NIST noise-parameter measurements," NIST Technical Note 1530 (March 2008). https://doi.org/10.6028/NIST.TN.1530 (Accessed 21 March, 2022).

27 J. Randa, "Detailed study of uncertainties in on-wafer transistor noise-parameter measurements," NIST Technical Note 1939 (October 2016). https://doi.org/10.6028/NIST.TN.1939 (Accessed 21 March, 2022).

6

Noise-Parameter Checks and Verification

6.1 Measurement of Passive or Previously Measured Devices

Because noise-parameter measurements are very challenging and can be derailed by many different small, subtle effects, it is desirable to have a means for checking that the results are correct, or at least not obviously incorrect. This is especially true if one is relatively inexperienced with such measurements or is using a new method or new instrumentation. Because verification methods rest on comparison of measurement results either to other measurement results or to theoretical predictions, they also serve as tests of the uncertainties ascribed to the measurements.

One check has already been discussed and demonstrated earlier, in Figure 5.9. That is the comparison of the value of the gain as measured in the noise measurements to the value obtained from vector network analyzer (VNA) measurements. While this is not a complete test, it is a good first check, and if the gain is determined in the noise measurements, it does not require extra measurements, since VNA measurements of the S-parameters of the device under test (DUT) are required for the noise-parameter measurements in any case.

One of the earliest verification methods for noise parameters or the noise figure was to measure the noise figure of an attenuator. If one writes Eq. (1.28) for an attenuator and forces it into the form $T_{out} = G(T_{in} + T_e)$, the result is that $T_e = (1/\alpha - 1)T_a$, where T_a is the noise temperature of the attenuator. Since α depends on the reflection coefficient of the input termination, so does T_e, as it should. This verification process is usually applied with a matched input load. Then, since attenuators are usually very well matched, $\alpha \approx |S_{21}|^2$, which can be identified as one over the loss L. (Note that $\alpha \leq 1, L \geq 1$.) If one makes the additional approximation that $T_a \approx T_0$, then the noise figure for the matched case becomes $F_0 \approx L$, $F_0(\mathrm{dB}) \approx L(\mathrm{dB})$. Thus, a 10 dB attenuator should have a noise figure of approximately 10 dB. If the gain is also measured in the noise measurements, it

Precision Measurement of Microwave Thermal Noise, First Edition. James Randa.

should yield a value of $G_0 = 1/L$. Because of the approximations that are usually involved, this is not a particularly stringent test of the uncertainties, but it is a relatively simple, basic check that there is not something grievously wrong.

Insofar as a typical attenuator is well matched, it is a good surrogate for most packaged amplifiers. However, because it is passive and does not have any gain, it does not present as much challenge to the measurement system as does an amplifier. Also, it does not present the difficulties associated with measuring a poorly matched amplifier or transistor. A poorly matched passive device can be used [1, 2] rather than an amplifier, but it still suffers from the lack of gain. An obvious choice for a test device, if it is available, would be an amplifier or transistor similar to the DUT(s) to be measured, but whose noise parameters are already known [3], e.g. from trusted measurements at some other laboratory. An active verification device has been designed and built for use in verification of Y-factor measurements of noise figure and cold-source methods [4].

Another check that was mentioned above, in Sections 4.4.2 and 5.4.2, is the measurement of T_{rev} [5, 6]. This requires a measurement of the noise temperature from the *input* of the DUT, when the output is terminated in a matched load, as in Figure 4.6. Using the measured noise parameters, one can predict the value of T_{rev} from Eq. (4.40) and compare that to the result of the direct measurement. Figure 6.1 shows an example of the results of this test [6] for an on-wafer transistor

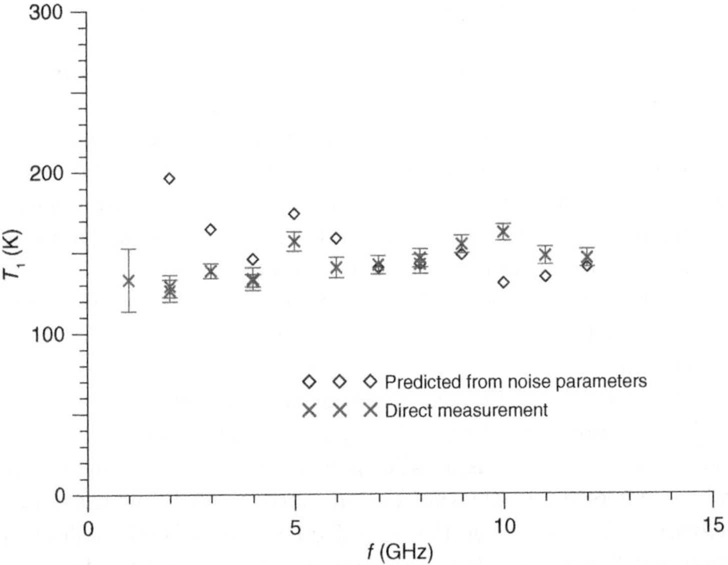

Figure 6.1 Application of the T_{rev} test, demonstrating problems at low frequencies. Source: Randa et al. [6]/IEEE.

at an industry laboratory. The disagreement of prediction and measurement at the lowest frequencies indicates a problem with either the measured noise parameters or the measurement of T_{rev}. The disadvantage of this check is that it requires an additional measurement in a configuration different from the other measurements.

6.2 Physical Bounds and Model Predictions

A relatively easy, but often not very stringent, test of noise-parameter measurement results is to check that the results do not violate various physical bounds. Such a check requires no additional measurements; at most it requires simple algebraic calculations using the measurement results. Some bounds are rather obvious, e.g. $T_{e,min} \geq 0$, $|\Gamma_{opt}| \leq 1$. Others can be less obvious, or are obvious only in a particular representation of the noise parameters. For example, the parameter X_1 in the wave representation is defined in Eq. (4.19) by $k_B X_1 = \langle |c_1|^2 \rangle$, which must be non-negative. If we use Eq. (4.21) to write X_1 in terms of the usual IEEE parameters, we obtain the requirement that

$$T_{e,min} \leq \frac{t|1 - S_{11}\Gamma_{opt}|^2}{\left(1 - |S_{11}|^2\right)|1 + \Gamma_{opt}|^2} \tag{6.1}$$

which is not at all obvious in the IEEE representation.

Similarly, the definitions of Eq. (4.19) require that

$$X_1 + X_2 \geq 2|X_{12}| \tag{6.2}$$

due to the Schwarz inequality. Even if the analysis is done in terms of the IEEE parameters, the wave-representation parameters can be computed from Eq. (4.21), and the resulting values must respect the bound of Eq. (6.2).

Another useful bound that we mention is $|\eta| \geq 2$, where

$$\eta = \frac{X_2\left(1 + |S_{11}|^2\right) + X_1 - 2Re\left(S_{11}^* X_{12}\right)}{(X_2 S_{11} - X_{12})} \tag{6.3}$$

This is a mathematical bound, derivable from the definition of η and Eq. (4.19). For well-matched amplifiers it is seldom violated, but for a poorly matched transistor it can be a useful check. Another general physical bound is that $T_{min} \leq 4R_n G_{opt} T_0$ [7, 8], where G_{opt} is the optimal conductance, the real part of Y_{opt} in Eq. (4.13).

Besides these bounds, which must be satisfied on the basis of fundamental mathematical or physical principles, there can be useful bounds that follow from transistor models. An example of such a bound is

$$2 \geq \frac{4R_n G_{opt} T_0}{T_{min}} \geq 1 \tag{6.4}$$

which is valid in a class of models for field effect transistors (FETs) and bipolar transistors [7, 9].

In applying any of these bounds, it should be remembered that they need to be satisfied only within the uncertainties. Thus, for example, a result of $T_{e,min} = -0.2 \pm 1.0\,\text{K}$ is consistent with the bound $T_{e,min} \geq 0$.

6.3 Tandem or Hybrid Measurements

If one is willing to perform additional measurements, beyond those required to just measure the noise parameters, then more powerful (but more complicated) verification methods are available. One such method uses an auxiliary passive two-port device. In this method, one measures the S-parameters of the auxiliary passive device (APD) and the DUT separately, measures the noise parameters of the DUT, and then attaches the auxiliary device to the DUT input and measures the S-parameters and noise parameters of the APD-DUT tandem configuration [10]. The three measurement configurations are shown in Figure 6.2. With the noise and S-parameters of the DUT measured, and with the noise parameters of the APD calculable from its measured S-parameters via Bosma's Theorem, one can also compute what the noise and S-parameters of the tandem configuration should be and compare those predictions to the values measured for the tandem configuration.

There are several attractive features of this verification method. One is that the verification measurements are performed on an active device (the tandem configuration), whose gain and noise figure can be comparable to those of the DUT. Another advantage is the flexibility that it affords – using different APDs allows different types of tests. Two APDs that have been successfully used are an isolator [11, 12] and a section of mismatched transmission line [13]. If an isolator is used as the auxiliary two-port, some of the noise parameters are approximately independent of the amplifier properties, and thus constitute a sort of independent absolute

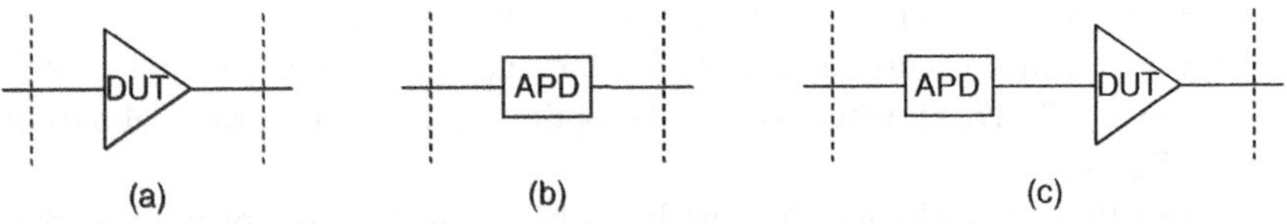

Figure 6.2 The three measurement configurations for the tandem test: (a) amplifier alone, (b) auxiliary passive device alone, and (c) tandem configuration.

standard, as opposed to just a consistency check. If a section of mismatched transmission line is chosen as the auxiliary two-port, the tandem configuration can be a highly reflective two-port, posing a more difficult measurement challenge. Furthermore, with an attenuator or a mismatched transmission line, the method can also be implemented in an on-wafer environment.

The appealing feature of using an isolator as the APD is that it leads to considerable simplifications that result in most of the predictions for the tandem configuration being independent of the measured noise parameters of the amplifier. It thus provides something close to an absolute reference standard, (most of) whose noise parameters can be calculated independent of any noise measurements. For example, the value for X_1 for the tandem configuration should be approximately equal to the isolator temperature. An even better test is provided by X_{12}. If the magnitude of X_{12} for the amplifier is considerably smaller than the isolator temperature, then

$$X'_{12} \approx -T_I S^I_{11} \tag{6.5}$$

where X'_{12} refers to the value for the tandem configuration, T_I is the isolator's noise temperature, and S^I_{11} is the value for the isolator.

Sample results of the tandem test using an isolator [10, 12] are shown in Figure 6.3. Only the results for X_{12} are shown, since they are the most interesting. (The other results were also successful.) The error bars correspond to the expanded uncertainties ($k = 2$), corresponding to confidence level of approximately 95%. The agreement is very good, which confirms (in this case) not only the noise-parameter measurement capability, but also the associated uncertainties. It is worth noting that the quantities being predicted and measured are quite small, from a few kelvins to a few tens of kelvins.

Thus, the isolator-amplifier combination provides something akin to a calculable standard, a device whose noise parameters (except X_2') are known (approximately) a priori from the ambient temperature and the isolator S-parameters. The isolated amplifier therefore provides an absolute test of the ability to measure the noise parameters correctly. Furthermore, with the (well matched) isolator one obtains a small value of X_{12}', which constitutes a very demanding test. A drawback of using an isolator as the passive device is that isolators are usually well matched, and consequently the tandem configuration does not provide a good test of the ability to measure poorly matched amplifiers. Also, an isolator would be difficult to implement in an on-wafer environment.

Poorly matched amplifiers present additional challenges, and the tandem test with a mismatched transmission line as the APD [10, 13] provides a consistency check of the ability to measure their noise parameters. Figure 6.4 shows some

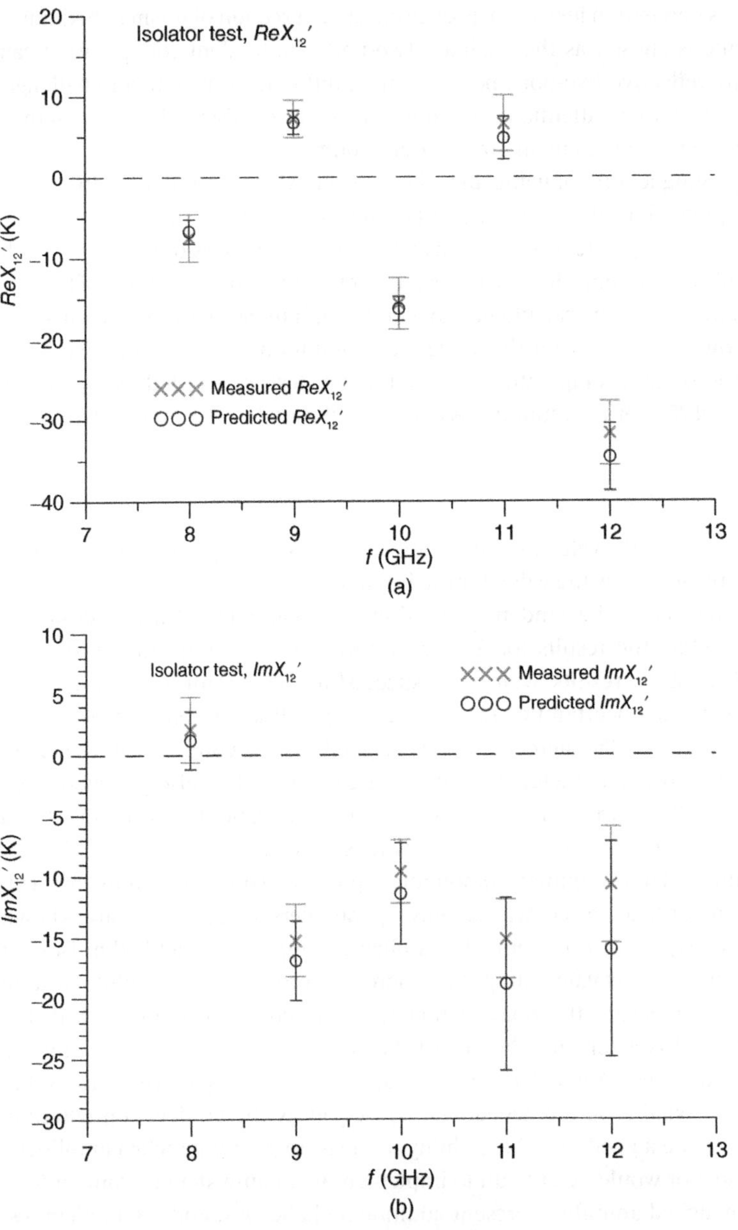

Figure 6.3 (a) Application of the tandem test with an isolator to ReX_{12} of an LNA. (b) Application of the tandem test with an isolator to ImX_{12} of an LNA. Source: Randa and Walker [12]/U.S. Department of Commerce/Public Domian.

Figure 6.4 Application of the tandem test with a mismatched transmission line to (a) F_{min}, (b) R_n, and (c) $|\Gamma_{opt}|$ of an LNA. Source: Randa et al. [10]/IEEE.

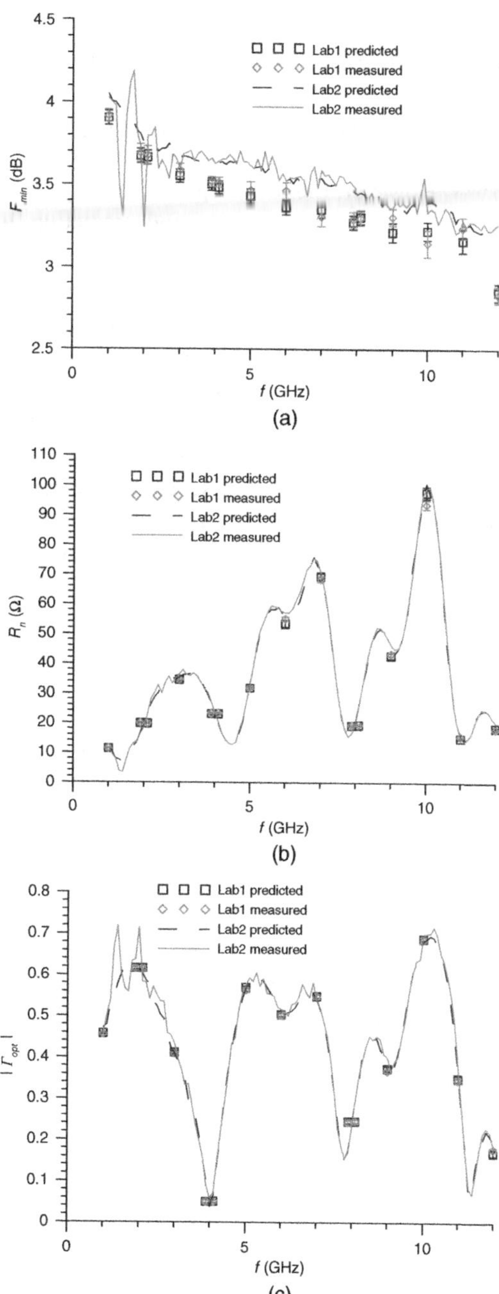

results from an application of this test to a particular amplifier [10]. These results demonstrate several different aspects of the tandem test with a mismatched transmission line. Most of the results for all three graphs and for both laboratories show good agreement between predicted and measured values for the tandem noise parameters, indicating no problems with the measurements for most cases. However, the low-frequency results for F_{min} for Lab 2 show a significant disagreement between theory and prediction, indicating a problem with either the measurements of the amplifier alone or the measurements of the tandem configuration (or both). There is also a less obvious discrepancy in the $|\Gamma_{opt}|$ results for Lab 2 at the same frequencies, Figure 6.4c. As it happens, these problems were not unexpected, because the system that was used for the measurements was intended for measurements of the 50 Ω noise figure, i.e. the noise figure for the case of a reflectionless input termination. It was not designed to measure noise parameters, but it can do so. Only four input states are used, most of which have relatively small $|\Gamma|$, and consequently the determination of noise parameters can be poor in the presence of large mismatch. Thus, Figure 6.4a,c demonstrate that the mismatched-line tandem test does indeed detect errors in measuring poorly matched amplifiers.

Figure 6.4a also demonstrates a limitation of the mismatched-line tandem test. Over most of the frequency range, the predicted and measured values of F_{min} agree for Lab 1 and for Lab 2, and yet there are significant differences between the Lab 1 and Lab 2 values; one or both of the sets of results are wrong. (In the absence of uncertainties for the Lab 2 results, it is impossible to be quantitative about the "significance" of the differences, but assuming that the Lab 2 uncertainties are not too large, it is rather obvious that there is a problem.) Thus, it is possible that a consistent measurement error will result in the verification test being satisfied despite incorrect results. An example of such an error would be an incorrect value for noise temperature of the hot input termination.

The tandem or hybrid test can also be implemented in on-wafer measurements [14], although it requires additional fabrication. Since isolators are not available on-wafer, the tandem test with isolators cannot be implemented on-wafer.

The properties of the various check and verification methods that have been discussed are summarized in Table 6.1.

Table 6.1 Summary of properties of different check and verification methods.

Method	Additional effort required	Consistency or absolute check	Applicable for matched or mismatched DUT	Applicable on-wafer?	Note
Passive-device measurement	Small	Absolute	Either	Yes	Does not test ability to measure active device
Measurement of known amplifier or transistor	Small	Absolute	Either	Yes	Requires device similar to DUT, with known properties
Gain comparison	Very small	Consistency	Either	Yes	
Physical bounds	Very small	Consistency	Either	Yes	Requires no additional measurements, just computation
T_{rev}	Significant	Consistency	Either	Yes	Requires measurement of reverse configuration
Tandem measurements-isolator	Significant	Absolute	Matched	No	
Tandem measurements-attenuator	Significant	Consistency	Matched	Yes	Use on-wafer requires additional fabrication
Tandem measurements-mismatched line	Significant	Consistency	Mismatched	Yes	Use on-wafer requires additional fabrication

References

1 A. Boudiaf, C. Dubon-Chevallier, and D. Pasquet, "Verification of on-wafer noise parameter measurements," *IEEE Transactions on Instrumentation and Measurement*, **44**, no. 2, pp. 332–335 (April 1995).

2 L. Escotte, R. Plana, J. Rayssac, O. Llopis, and J. Graffeuil, "Using cold FET to check accuracy of microwave noise parameter test set," *Electronics Letters*, **27**, no. 10, pp. 833–835 (May 1991).

3 C.A. Morales-Silva, L. Dunleavy, and R. Connick, "Noise parameter measurement verification by means of benchmark transistors," *High Frequency Electronics*, pp. 18–25, (February 2009).

4 J. Dunsmore, "Noise figure verification of Y-factor and cold source methods," in *2017 International Conference on Noise and Fluctuations (ICNF)*, 2017, pp. 1–4. doi: 10.1109/ICNF.2017.7985945.

5 D. Wait and G.F. Engen, "Application of radiometry to the accurate measurement of amplifier noise," *IEEE Transactions on Instrumentation and Measurement*, **40**, no. 2, pp. 433–437 (April 1991).

6 J. Randa, T. McKay, S.L. Sweeney, D.K. Walker, L. Wagner, D.R. Greenberg, J. Tao, and G.A. Rezvani, "Reverse noise measurement and use in device characterization," in *2006 IEEE Radio Frequency Integrated Circuits (RFIC) Symposium Digest*, June 2006, pp. 345–348. doi: 10.1109/RFIC.2006.1651152

7 M.W. Pospieszalski, "Interpreting transistor noise," *IEEE Microwave Magazine*, **11**, no. 6, pp. 61–69 (October 2010).

8 M.W. Pospieszalski and W.Wiatr, "Comment on 'Design of microwave GaAs MESFET's for broadband, low-noise amplifiers'," *IEEE Transactions on Microwave Theory and Techniques*, **MTT-34**, no. 1, p. 194 (January 1986).

9 M.W. Pospieszalski, "On certain noise properties of field-effect and bipolar transistors," in *Proceedings of Microwaves, Radar and Wireless Communication ((MIKON) 2006 Conference*, Vol. 3, Krakow, Poland, May 2006, pp. 1127–1130.

10 J. Randa, J. Dunsmore, D. Gu, K. Wong, D.K. Walker, and R.D. Pollard, "Verification of noise-parameter measurements and uncertainties," *IEEE Transactions on Instrumentation and Measurement*, **60**, no. 11, pp. 3685–3693 (November 2011). doi: 10.1109/TIM.2011.2138270

11 M. Pospieszalski, "On the noise parameters of isolator and receiver with isolator at the input," *IEEE Transactions on Microwave Theory and Techniques*, **MTT-34**, no. 4, pp. 451–453 (April 1986).

12 J. Randa and D.K. Walker, "Amplifier noise-parameter measurement checks and verification," in *63rd ARFTG Conference Digest*, Ft. Worth, TX, June 2004, pp. 41–45. doi: 10.1109/ARFTG.2004.1387853.

13 K. Wong, R. Pollard, B. Shoulders, and L. Rhymes, "Using a mismatch transmission line to verify accuracy of a high performance noise figure measurement system," in *69th ARFTG Conference Digest*; Honolulu, HI, June 2007, pp. 170–174.

14 A. Wu, X. Fu, C. Liu, C. Li Y. Wang, F. Liang, and P. Luan, "Optimal design of passive devices for verifying on-wafer noise parameter measurement systems," *IEEE Transactions on Instrumentation and Measurement*, **69**, no. 6, pp. 2837–2844 (June 2020).

7

Cryogenic Amplifiers

7.1 Background

7.1.1 Introduction

Not surprisingly, the amplifiers that achieve the lowest effective input noise temperatures are those that operate at cryogenic temperatures. Because of the inconvenience of maintaining cryogenic conditions, cryogenic amplifiers are only used when the very lowest levels of added noise are required. The traditional application for cryogenic amplifiers is in radio astronomy [1, 2]. They are also used in mm-wave and terahertz imaging [3], and recently there has been a surge of interest spurred by the development of quantum computing [4]. A recent review can be found in [5]. In all applications, the low noise temperatures achieved by cryogenic amplifiers require that their measurement be particularly accurate.

7.1.2 Vacuum-Fluctuation Contribution

When discussing precision measurements of cryogenic amplifiers, it becomes necessary to include the effect of vacuum fluctuations. This is a quantum mechanical effect arising from the presence of a sea of virtual particle-antiparticle pairs that permeates the vacuum everywhere and at all times. Because the vacuum is the lowest possible energy state, energy conservation prevents any net energy from being extracted from this sea. It can, however, affect physical processes, and in particular, it adds noise to active electronic devices [6–8]. We have neglected mentioning this effect earlier because it is entirely negligible for room-temperature devices and measurements. For cryogenic amplifiers, however, the noise added by the amplifier can be so small that the vacuum-fluctuation contribution can be (barely) noticeable at microwave frequencies, and we need to consider it in the definition of the effective input noise temperature T_e.

Precision Measurement of Microwave Thermal Noise, First Edition. James Randa.

The vacuum fluctuations result in an additional effective input noise temperature of

$$T_{vac} = \frac{hf}{2k_B} \tag{7.1}$$

at the input of an amplifier. This results in a minimum output noise from an amplifier, $N_{out,min} = Ghf/2$. The effect is very small and usually negligible at microwave frequencies; for example, $T_{vac} = 0.24$ K at 10 GHz. However, there are cases, such as very low-noise cryogenic amplifiers [9, 10], where it is not negligible and must be taken into account.

There is not yet general agreement on how to include T_{vac} in the definition of noise temperatures when dealing with microwave amplifiers [9, 11, 12]. If a noise source R is connected to the amplifier input, the output noise temperature will be $T_{out} = G(T_R + T_{vac} + T_e)$, as shown in Figure 7.1, but should T_{vac} be grouped with T_R or with T_e? I.e. should it be considered as part of the input noise or as additional noise added by the amplifier? We will to treat T_{vac} as part of the input noise, since it is due to the environment in which the amplifier operates rather than to the amplifier itself. Thus, it does not contribute to T_e, the noise added by the amplifier, nor does it contribute to the noise figure. We also choose to consider it as an additional input source, rather than as a modification of T_R, the noise temperature of the input noise source, Figure 7.1b. This choice is motivated in part by the possibility of a large separation distance between the input source and the amplifier, as for example occurs in remote sensing. In such a case, it seems more natural to ascribe the effect to the vacuum that occurs at the amplifier input, rather than the vacuum present at the distant source.

Thus our equation relating the output to the input noise temperature for a cryogenic amplifier is

$$T_{out} = G_{av}(T_{in} + T_e + T_{vac}) \tag{7.2}$$

where T_{in} is the noise temperature of the input source, as defined in Chapter 1, T_e is the effective input noise temperature of the amplifier, and $T_{vac} = hf/(2k_B)$ is the

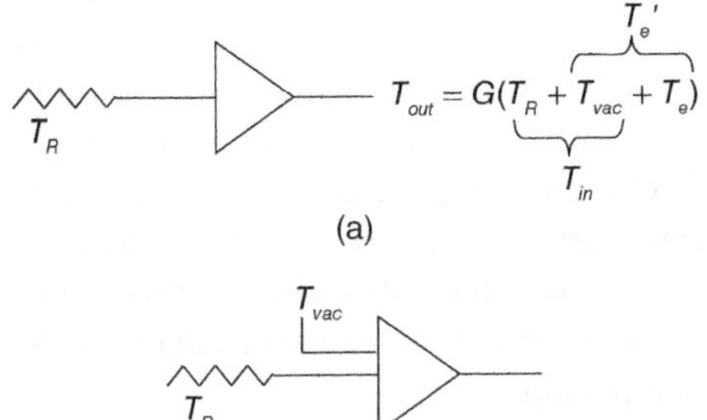

Figure 7.1 (a) Two options for treatment of T_{vac}. (b) Vacuum contribution as a separate input noise source.

vacuum-fluctuation contribution. This definition would then also carry over to a treatment of the noise parameters of a cryogenic amplifier, where T_{vac} would be considered a separate contribution to the input noise.

7.2 Measurement of the Matched Noise Figure

7.2.1 Cold-Attenuator Method

The fundamental problem in measuring the noise figure of a cryogenic amplifier is that the input and output ports of the amplifier, planes 1 and 2 in Figure 7.2, are not readily accessible. A simple application of the Y-factor method would measure the noise figure between planes 0 and 3, which would include the effects of the connecting lines. Consequently, one must somehow characterize and correct for the effects of those lines in order to obtain a good measurement of the desired noise figure of the amplifier between planes 1 and 2. (In this section, we use "the noise figure" to refer to the noise figure for matched input sources.)

The usual method for measuring the noise figure of a cryogenic amplifier is the cold-attenuator method. It was originated by Anthony Kerr in the 1970s and was first published by Sander Weinreb [13]. A good explanation of it can be found in [14]. We assume the input noise source is a noise diode, switched between its on and off states. The basic idea is to introduce an attenuator in front of the amplifier, within the cryostat, as in Figure 7.3a. (It is assumed that the receiver-power meter combination has already been calibrated.) This results in two significant improvements. Most importantly, it reduces the dependence on any effect of the input transmission line. When the diode is in its off state, the noise temperature at the input to the amplifier is very well known (and quite low), with little dependence on any losses in the input transmission line. Even in the on state the effect of the input line is reduced. In addition, the effect of any difference in the reflection coefficient of the diode between its on and off states is greatly reduced.

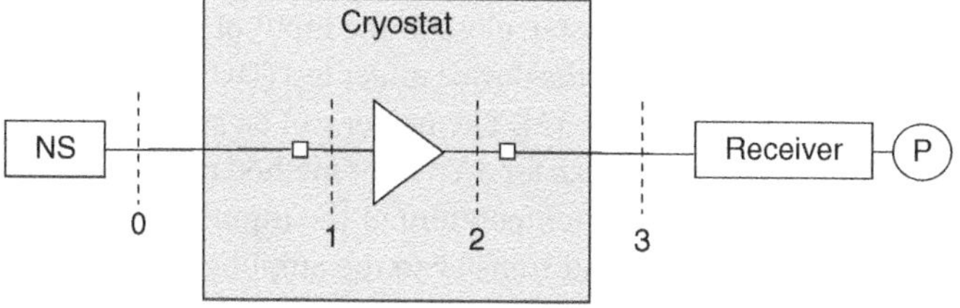

Figure 7.2 Reference planes for noise-figure measurement on a cryogenic amplifier. The small blocks near planes 1 and 2 represent the location of the internal connectors relative to those planes.

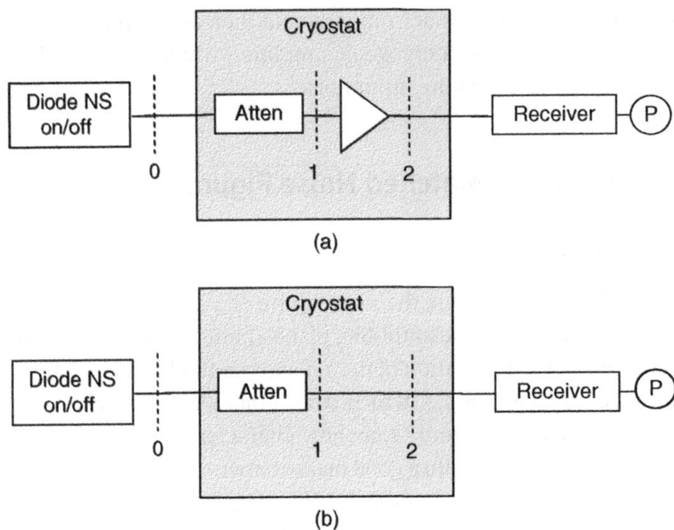

Figure 7.3 (a) Noise-figure measurement for cryogenic amplifier in the cold-attenuator method. (b) Supplemental configuration for characterizing lines in the cold-attenuator method.

The cold-attenuator method can be refined by performing additional measurements to characterize the input and output transmission lines and the attenuator at the cryogenic temperature, as indicated in Figure 7.3b. The authors of [14, 15] performed measurements with and without the attenuator in Figure 7.3b. Furthermore, they took considerable care to ensure that input and output lines are symmetrical, so that any measured losses could be apportioned equally between input and output. In this manner, they were able to achieve uncertainties in the noise figure a little above 1 K.

7.2.2 Internal Hot–Cold Method

The difficulty in characterizing the input transmission line can be avoided by locating the input noise source within the cryostat, close to the input of the amplifier. A matched load can be positioned close to the device under test (DUT) input, with a heater attached to the matched load [14, 15]. The heater can be switched on or off to provide hot or cold input noise sources for a Y-factor method measurement. A thermometer is needed to monitor the temperature of the input noise source. Some care must be taken to minimize heat transfer to the amplifier input when the heater is on. In principle this procedure provides hot and cold input sources that are known quite well at the input measurement plane. It does not, however, account for the effect of the output transmission line (from plane 2 to plane 3 in Figure 7.2). There is also a question of how well matched the input noise sources are at the two input temperatures.

7.2.3 Full-Characterization Measurements

For a particular, (very) low-noise amplifier (LNA), uncertainties a little above 1 K in the effective input noise temperature of the amplifier were attained with the cold-attenuator method [14, 15]. It is possible to improve that uncertainty, but (considerably) more measurement (and some analytic) effort is required.

Referring to Figure 7.2, the following basic noise equations govern the relations between the noise temperatures at the various reference planes:

$$T_2 = G(T_1 + T_e + T_{vac})$$

$$T_3 = \alpha_{32} T_2 + \Delta T_2$$

$$T_1 = \alpha_{10} T_{in} + \Delta T_1' \tag{7.3}$$

where ΔT_2 is the cumulative noise added by the line in going from the cryogenic temperature at plane 2 to room temperature at plane 3. On the other hand, $\Delta T_1{}'$ is the noise added going from room temperature at plane 0 to the cryogenic temperature at plane 1. We adopt the convention that the unprimed quantities ΔT_1 and ΔT_2 will be used for the direction from inside the cryostat to the outside (plane 1 to plane 0, and 2 to 3), whereas the primed quantities $\Delta T_1{}'$ and $\Delta T_2{}'$ will be used for the opposite direction, from outside in.

The fact that there is directional dependence to these quantities may seem counterintuitive at first, but it becomes less so if one considers the case of two attenuators at different temperatures and computes the output noise temperature in the two directions for the same input noise temperature. (It is similar to computing the noise temperature of cascaded amplifiers for different ordering of the amplifiers.) The ordering of the attenuators does make a difference.

If we combine the three equations of Eq. (7.3), we obtain

$$T_3 = \alpha_{10}\alpha_{32} G T_{in} + \alpha_{32} G \Delta T_1' + \alpha_{32} G(T_e + T_{vac}) + \Delta T_2 \tag{7.4}$$

for the output noise temperature in Figure 7.2. Using two different input noise sources T_{in}, we can measure T_3 for each and thus use Eq. (7.4) to determine $\alpha_{10}\alpha_{32}G$. However, to determine G and T_e separately, we need to also determine $\alpha_{10}\alpha_{32}$, α_{32}, $\Delta T_1{}'$, and ΔT_2. They can be determined by the series of auxiliary measurements shown in Figure 7.4.

The configuration of Figure 7.4a, where a through section is used between planes 1 and 2, is used with two different input noise sources to determine $\alpha_{10}\alpha_{32}$ by the usual Y-factor method. Combined with the measurements in Figure 7.2, this yields a measurement of the amplifier gain G.

In Figure 7.4b, an internal matched load (at liquid-helium temperature, of course) is connected at plane 2, which provides a measurement of ΔT_2. (It also provides a check of whether the matched load remains matched at cryogenic temperature.) Similarly, the configuration in Figure 7.4c provides

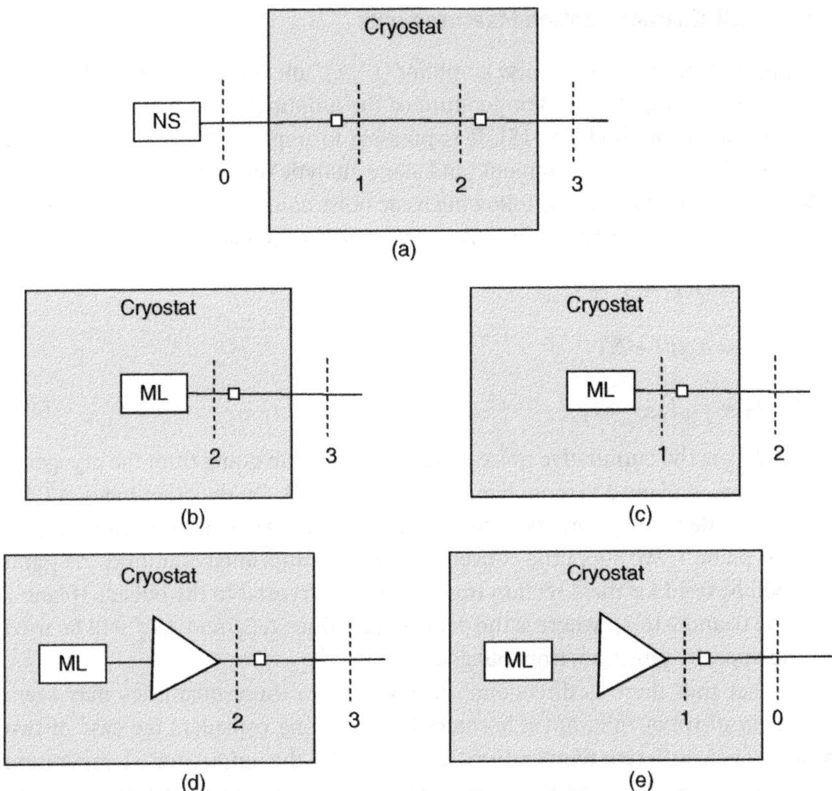

Figure 7.4 Auxiliary measurements used to determine α_{10}, α_{32}, α_{32}, $\Delta T_1{}'$, and ΔT_2. (a) Through measurement. (b) Cold load at internal output plane. (c) Cold load at internal input plane. (d) Amplifier at internal output plane. (e) Amplifier at internal input plane. (Source: Randa et al. [9]/IEEE.)

a measurement of ΔT_1. The configurations of Figure 7.4d,e exploit the cryogenic environment by using a cryogenic-temperature input noise source. When combined with the results of Figures 7.2 and 7.4a–c (and some analysis [9]), this enables us to determine G and T_e. In relating the quantities of interest ($\alpha_{10}\alpha_{32}$, α_{32}, $\Delta T_1{}'$, and ΔT_2) to the measured outputs in Figure 7.4, considerable care must be exercised to account for directional dependence of the available power ratios (the α's) and the Δ's. That care is exercised in [9] and will not be repeated here.

Using this method and the associated uncertainty analysis, the authors of [9] were able to measure values of T_e as low as 2.3 K with a standard uncertainty of 0.3 K. Their results for G and T_e are shown in Figure 7.5. In subsequent measurements [10] on a different cryogenic amplifier, this method yielded noise

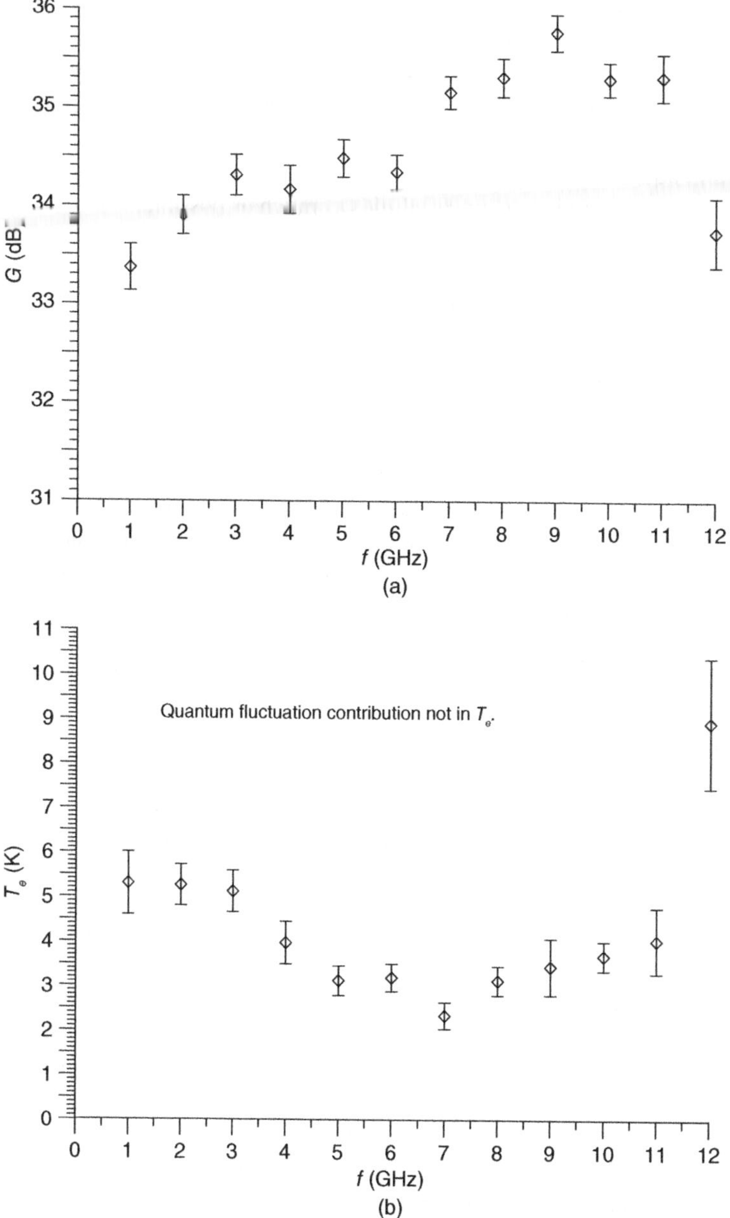

Figure 7.5 (a) Measured gain of a particular cryogenic amplifier. (b) Measured effective input noise temperature of a particular cryogenic amplifier. Source: Randa et al. [9]/IEEE.

temperatures as low as 1.65 ± 0.18 K. A method was also developed and employed to verify the results and uncertainties, and the checks were satisfied [16].

Although the characterization of the lines yielded significant improvement in the uncertainties, this improvement came at a high price in measurement time, since multiple additional cool-down cycles were required. The added costs in time and effort (and liquid helium) mean that this not a practical method unless one is highly motivated to minimize the uncertainties as much as possible.

7.3 Noise-Parameter Measurement

As with room-temperature devices, noise-figure measurements yield useful indicators of the overall noise performance of a cryogenic amplifier or transistor, but if one is to design circuits for optimal noise performance, it is necessary to know not only the amplifier's or transistor's matched noise figure, but also the noise parameters that specify the noise performance as a function of the reflection coefficient of the input termination. Accurate measurement of noise parameters (cf. Chapter 5) is challenging under the best of conditions; cryogenic amplifiers operate (and must be measured) under conditions that introduce additional challenges. Nonetheless, applications such as radioastronomy, quantum computing, and communication systems employ cryogenic LNAs, which must then be (noise) matched to surrounding circuitry.

The conventional procedure for measuring room-temperature amplifier or transistor noise parameters depends on measuring the output for a series of different input reflection coefficients. For a cryogenic device the relevant input reference plane is within the cryostat, at the device input (plane 1 in Figure 7.2). Providing different input reflection coefficients at plane 1 is (conceptually) relatively simple, either by a tuner or discrete terminations outside the cryostat (plane 0 in Figure 7.2) or by a tuner or discrete terminations within the cryostat, directly at plane 1. If the tuner or discrete terminations are located outside the cryostat, however, the effects of the transmission line connecting the room-temperature input terminations to the cryogenic amplifier input must be included. These include the effects on both the input reflection coefficient and on the input noise temperature. Furthermore, as with the probes in on-wafer measurements, losses in the transmission line will restrict the range of input reflection coefficients available to the amplifier. If, on the other hand, the tuner or discrete terminations are to be located within the cryostat, they must be compact and must operate at cryogenic temperatures. This usually requires hardware to be specially designed and built for that specific application.

Early work on cryogenic noise-parameters [13, 17] included using variable impedances at both the input and the output of the DUT [13], where the impedances could be varied manually from outside the cryostat. In [18–20] an input impedance tuner was located external to the cryocooler, but the coaxial-to-wafer transition was located within the cryocooler, and an on-wafer calibration was performed at cryogenic temperature. In [5], an impedance tuner was located inside the cryostat, at the DUT input. The tuner was designed to generate a minimal set of four input impedances [21] as discussed in Section 4.5.5.

Weinreb and collaborators have designed and employed a "long-line module" for use within the cryostat [22, 23]. As the name implies, this module includes a length of transmission line to be used in conjunction with the frequency-variation method discussed above, in Section 4.5.7, as well as a noise diode and a heated plate (with temperature sensor) to vary and monitor the physical temperature of the unit. An external diode source is also used. The use of the heated plate essentially provides a (variable) primary noise-temperature standard right at the DUT input as in the Internal Hot–Cold Method of Section 7.2.2, thus evading the problem of noise added by the transition from room temperature to cryogenic temperature.

One residual concern in assessing the measurement uncertainties in all these approaches is the question of how well the source reflection coefficient is really known at the input reference plane.

References

1 J.C. Weber and M.W. Pospieszalski, "Microwave instrumentation for radio astronomy," *IEEE Transactions on Microwave Theory and Techniques,* **50**, no. 3, pp. 986–995 (March 2002).

2 M.W. Pospieszalski, "Extremely low-noise amplification with cryogenic FET's and HFET's: 1970–2004," National Radio Astronomy Observatory, Charlottesville, VA, Electronics Division Internal Report 314 (May 2005). Available: http://www.gb.nrao.edu/electronics/edir/edir314.pdf (Accessed 14 March, 2022).

3 F. Rodriguez-Morales, K.S. Yngvesson, R. Zannoni, E. Gerecht, D. Gu, X Zhao, N. Wadefalk, and J.J. Nicholson, "Development of integrated HEB/MMIC receivers for near-range terahertz imaging, *IEEE Transactions on Microwave Theory and Techniques,* **54**, pp. 2301–2311 (2006).

4 B. Patra, R.M. Incandela, J.P.G. van Dijk, H.A.R. Homulle, L. Song, M. Shahmohammadi, R.B. Staszewski, A. Vladimirescu, M. Babaie, F. Sebastiano, and E. Charbon, "Cryo-CMOS circuits and systems for quantum computing applications," *IEEE Journal of Solid-State Circuits,* **53**, no. 1, pp. 309–321 (January 2018). doi: 10.1109/JSSC.2017.2737549.

5 A. Sheldon, L. Belostotski, H. Mani, C.E. Groppi and K.F. Warnick, "Cryogenic noise-parameter measurements: recent research and a fully automated measurement application," *IEEE Microwave Magazine*, **22**, no. 8, pp. 52–64 (August 2021). doi: 10.1109/MMM.2021.3078027.

6 H.B. Callen and T.A. Welton, "Irreversibility and generalized noise," *Physics Review*, **83**, no. 1, pp. 34–40. (July 1951).

7 S.M. Caves, "Quantum limits on noise in linear amplifiers," *Physical Review D*, Ser. 3, **26**, no. 8, pp. 1817–1839 (October 1982).

8 J.R. Tucker and M.J. Feldman, "Quantum detection at millimeter wavelengths," *Reviews of Modern Physics*, **57**, no. 4, pp. 1055–1113 (October 1985).

9 J. Randa, E. Gerecht, D. Gu, and R.L. Billinger, "Precision measurement method for cryogenic amplifier noise temperatures below 5 K," *IEEE Transactions on Microwave Theory and Techniques*, **54**, no. 3, pp. 1180–1189 (March 2006). doi: 10.1109/TMTT.2005.864107.

10 D. Gu, J. Randa, R. Billinger, and D.K. Walker, "Measurement and uncertainty analysis of a cryogenic low-noise amplifier with noise temperature below 2 K," *Radio Science*, **48**, pp. 344–351 (2013). doi: 10.1002/rds.20039.

11 A.R. Kerr, "Suggestions for revised definitions of noise quantities, including quantum effects," *IEEE Transactions on Microwave Theory and Techniques*, **47**, pp.325–329 (March 1999).

12 A.R. Kerr and J. Randa, "Thermal noise and noise measurements—a 2010 update," *IEEE Microwave Magazine*, **11**, no. 6, pp. 40–52 (October 2010). doi: 10.11009/MMM.2010.937732

13 S. Weinreb, "Low-noise cooled GASFET amplifiers," *IEEE Transactions on Microwave Theory and Techniques*, **MTT-28**, no. 10, pp. 1041–1054 (1980).

14 J.E. Fernandez, "A noise-temperature measurement system using a cryogenic attenuator," TMO Progress Report 42-135 (November 1998). https://tmo.jpl.nasa.gov/progress_report/42-135/135F.pdf (Accessed 12 March, 2022).

15 N. Wadefalk, A. Mellberg, I. Angelov, M.E. Barsky, S. Bui, E. Choumas, R.W. Grundbacher, E.L. Kollberg, R. Lai, N. Rorsman, P. Starski, J. Stenarson, D.C. Streit, and H. Zirath, "Cryogenic wide-band ultra-low-noise IF amplifiers operating at ultra-low DC power," *IEEE Transactions on Microwave Theory and Techniques*, **51**, no. 6, pp. 1705–1711 (June 2003).

16 D. Gu, J. Randa, R. Billinger, and D.K. Walker, "A verification method for noise-temperature measurements on cryogenic low-noise amplifiers," in *2012 Conference on Precision Electromagnetic Measurements (CPEM 2012) Conference Digest*, Washington, DC, July 2012, pp. 32–33. doi: 10.1109/CPEM.2012.6250644.

17 M. Pospieszalski, "On the measurement of noise parameters of microwave two-ports," *IEEE Transactions on Microwave Theory and Techniques*, **MTT-34**, no. 4, pp. 456–458 (April 1986).

18 L. Escotte, F. Sejalon and J. Graffeuil, "Noise parameter measurement of microwave transistors at cryogenic temperature," *IEEE Transactions on*

Instrumentation and Measurement, **43**, no. 4, pp. 536–543 (August 1994). doi: 10.1109/19.310165.

19 I. Rolfes, T. Musch, and B. Schiek, "Cryogenic noise parameter measurements of microwave devices," *IEEE Transactions on Instrumentation and Measurement*, **50**, no. 2, pp. 373–376 (April 2001).

20 W. Wiatr, "Comments on "Cryogenic noise parameter measurements of microwave devices"," *IEEE Transactions on Instrumentation and Measurement*, **53**, no. 2, p. 619 (April 2004). doi: 10.1109/TIM.2004.823649.

21 M. Himmelfarb and L. Belostotski, "On impedance-pattern selection for noise parameter measurement," *IEEE Transactions on Microwave Theory and Techniques*, **64**, no. 1, pp. 258–270 (January 2016). doi: 10.1109/TMTT.2015.2504500.

22 R. Hu and S. Weinreb, "A novel wide-band noise-parameter measurement method and its cryogenic application," *IEEE Transactions on Microwave Theory and Techniques*, **52**, no. 5, pp. 1498–1507 (May 2004). doi: 10.1109/TMTT.2004.827029.

23 D. Russell and S. Weinreb, "Cryogenic self-calibrating noise parameter measurement system," *IEEE Transactions on Microwave Theory and Techniques*, **60**, no. 5, pp. 1456–1467 (May 2012). doi: 10.1109/TMTT.2012.2188813.

8

Multiport Amplifiers

8.1 Introduction

We have discussed the noise figure (or the effective input noise temperature) for two-port amplifiers at some length in the preceding chapters. In general, the noise figure is not an intrinsic property of the device, but rather it depends on the reflection coefficients of the terminations. That dependence can be characterized by a set of noise parameters, and there are several equivalent parameterizations. These include the familiar IEEE set and its variants as well as the noise matrix, either in its voltage–current form or its wave-amplitude incarnation.

Real-world applications, however, include amplifiers with more than one input and one output port, such as differential amplifiers in cell phones (and elsewhere) and feeds for antenna arrays. For more than two ports, or for more than one mode in a single port, the situation is not so well developed. The basic multiport noise-matrix formalism was introduced long ago [1], but a formalism for noise figures and noise parameters was not developed or established. Even the definition of multiport noise figures is not well established. The IEEE definitions [2] allow for multiple input ports, as well as different input and output frequencies (since they were developed with receivers in mind), but they are restricted to one output port and, even for that case, they stop short of defining a noise figure, much less noise parameters. Circuit analysis and scattering parameters have been developed for differential and common modes in mixed-mode two-ports (i.e. two *physical* ports) [3], but for general multiport noise we find it more natural to use a framework based on the wave representation of the noise matrix [4–7]. This chapter is based on the development in [6].

Precision Measurement of Microwave Thermal Noise, First Edition. James Randa.

8.2 Formalism and Noise Matrix

A single port will be taken to be a single mode in a single physical port. For purposes of this discussion, different modes in a single physical port will be considered to be different ports. The analysis is done in terms of wave amplitudes, which may be defined in terms of voltages and currents [3, 8, 9], or they may be introduced and used with no reference to voltage and current [10, 11].

Details of the modes or waves are not of concern, except for two general properties. First, the modes must be power-orthogonal, i.e. the total power across a reference plane must be the sum of the powers in each of the individual modes or ports. If this is not the case, and the expression for the total power contains cross terms among the modes, it is possible (at least for lossless or low-loss lines) to achieve power orthogonality by means of a linear transformation [12]. The second condition on the waves is that they can be physically generated in practical applications, otherwise the discussion of measurements based on these waves is purely academic.

The notation is outlined in Figure 8.1. Generalizing the two-port case of Chapter 1, a linear M-port amplifier can be represented by its $M \times M$ scattering matrix ($\boldsymbol{S}$) and an M-vector of internal noise sources ($\boldsymbol{c}$),

$$\boldsymbol{b} = \boldsymbol{S}\boldsymbol{a} + \boldsymbol{c} \tag{8.1}$$

where $\boldsymbol{a}$ and $\boldsymbol{b}$ are M vectors of the usual incident and outgoing wave amplitudes, and $\boldsymbol{c}$ is the M vector whose ith element, c_i is the wave amplitude of the generator wave emitted from port i due to sources internal to the M-port amplifier. The amplitude c_i would be the output noise amplitude for reflectionless terminations and no input noise. The incident noise wave $\boldsymbol{a}$ can be written as

$$\boldsymbol{a} = \boldsymbol{\Gamma}\boldsymbol{b} + \hat{\boldsymbol{a}}, \tag{8.2}$$

where $\hat{\boldsymbol{a}}$ is the vector of generator waves of the sources connected to the M ports, and $\boldsymbol{\Gamma}$ is the $M \times M$ matrix of reflection coefficients of those sources. In simple

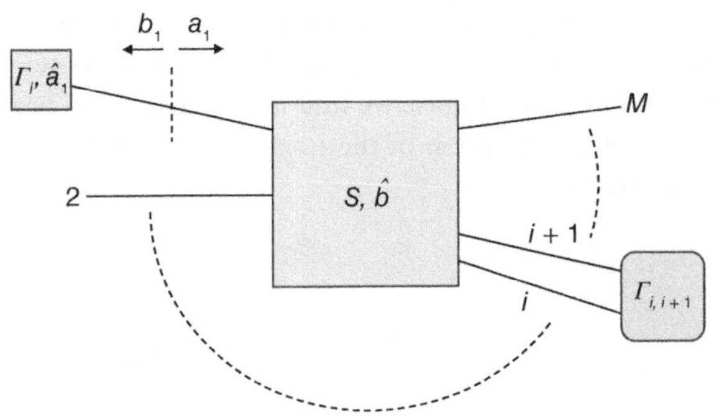

Figure 8.1 Illustration of notation. Source: Randa [6]/IEEE.

cases, $\boldsymbol{\Gamma}$ is diagonal and Γ_{ii} is the reflection coefficient from the termination on port i. More generally, $\boldsymbol{\Gamma}$ has off-diagonal elements Γ_{ij} corresponding to a wave emerging from port or mode j and being reflected back (at least partially) in port or mode i.

Combining Eqs. (8.1) and (8.2) in the usual manner yields the expression for the outgoing wave vector in terms of the generator waves,

$$\boldsymbol{b} = [\mathbf{1} - \boldsymbol{S\Gamma}]^{-1}[\boldsymbol{S}\hat{\boldsymbol{a}} + \boldsymbol{c}] \tag{8.3}$$

One can then define the noise matrix as in Section 1.2.2 and express it in terms of the intrinsic properties of the M port and the properties of the terminations on its ports. The full, in situ noise matrix is defined as

$$\boldsymbol{N} \equiv \langle \boldsymbol{b}\boldsymbol{b}^{\dagger} \rangle \tag{8.4}$$

or

$$N_{ij} = \left\langle b_i b_j^* \right\rangle \tag{8.5}$$

where the brackets indicate an ensemble or time average (assumed to be the same), and the † indicates Hermitian conjugate.

Equation (8.3) can then be used to write the noise matrix in terms of the generator waves $\hat{\boldsymbol{a}}$ and $\boldsymbol{c}$ as

$$\boldsymbol{N} = [\mathbf{1} - \boldsymbol{S\Gamma}]^{-1}[\boldsymbol{S}\langle \hat{\boldsymbol{a}}\hat{\boldsymbol{a}}^{\dagger} \rangle \boldsymbol{S}^{\dagger} + \langle \boldsymbol{c}\boldsymbol{c}^{\dagger} \rangle][\mathbf{1} - \boldsymbol{S\Gamma}]^{-1\dagger} \tag{8.6}$$

where we have used the fact that the generator waves from the amplifier are uncorrelated with those from the terminations. The term involving $\hat{\boldsymbol{a}}\hat{\boldsymbol{a}}^{\dagger}$ is due to the noise (and signals) from the sources and terminations connected to the ports, and the $\boldsymbol{c}\boldsymbol{c}^{\dagger}$ term is due to the noise generated by the amplifier itself, suitably modified by reflections from the terminations.

The in situ noise matrix describes the noise behavior of the M-port amplifier embedded in surrounding circuitry whereas the noise properties of the M port itself are embodied in the *intrinsic* noise matrix, defined by

$$\hat{\boldsymbol{N}} = \langle \boldsymbol{c}\boldsymbol{c}^{\dagger} \rangle \tag{8.7}$$

The intrinsic noise matrix, supplemented by the S matrix, contains full information about the intrinsic noise properties of the device. For further discussion, it is useful to introduce a more physical parameterization of it, as well as more compact notation. For diagonal elements of the intrinsic noise matrix, we define a noise temperature for each port,

$$\left\langle |c_i|^2 \right\rangle = k_B \hat{T}_i \tag{8.8}$$

The quantity $k_B\hat{T}_i$ is the noise power per unit bandwidth that would be delivered to a noiseless, reflectionless load attached to port i if all other ports were also terminated in noiseless reflectionless loads. The actual noise temperature of port i is related to the *available* noise power, and for the case of all noiseless reflectionless terminations, the noise temperature is given by $T_{i,0} = \widehat{T}_i / \left(1 - |S_{ii}|^2\right)$. Off-diagonal elements of the intrinsic noise matrix are appropriately scaled correlation functions,

$$\left\langle c_i c_j^* \right\rangle = k_B \sqrt{\widehat{T}_i \widehat{T}_j} \rho_{ij} \tag{8.9}$$

The absolute values of the ρ_{ij} fall between 0 and 1, as befits a correlation function. For the discussion that follows, it is also useful to define a correlation matrix for the incident noise waves. Let

$$\boldsymbol{A} \equiv \left\langle \widehat{a}_i \widehat{a}_j^* \right\rangle = k_B T_0 A_{ij} \tag{8.10}$$

Like $\boldsymbol{\Gamma}$, $\boldsymbol{A}$ is not necessarily diagonal, since terminations of various ports may be interconnected. Diagonal elements of $\boldsymbol{A}$ are ratios of the incident noise temperatures to T_0.

With multiple ports, there may be several different useful choices for the set of basis waves (e.g. common and differential modes of ports 1 and 2). The noise matrix (in situ or intrinsic) will be different for different choices of basis waves. Under a change of basis represented by the matrix $\boldsymbol{L}$, $\boldsymbol{b} \rightarrow \boldsymbol{b}' = \boldsymbol{L}\boldsymbol{b}$, the noise matrix transforms according to

$$\boldsymbol{N} \rightarrow \boldsymbol{N}' = \boldsymbol{L}\boldsymbol{N}\boldsymbol{L}^{-1} \tag{8.11}$$

8.3 Definition of Noise Figure for Multiports

We now address the issue of the definition of the noise figure and its expression in terms of the intrinsic parameters of the amplifier and its terminations. We deal only with the noise figure at a single frequency, known as the spot noise figure. We saw in Chapter 4 that for a two port, the noise figure at a given frequency is defined as the ratio of total output noise power per unit bandwidth to the portion of the output noise power that is due to the input noise, evaluated for the case where the input noise power is $k_B T_0$. So, how do we extend that to the multiport case? In principle, we could define noise figures for all the ports, but in practice we will consider noise figures only for output ports.

As in the two-port case, the noise figure of a given port should be the ratio of the total output noise power in that port to the output noise power that is due to the input noise, for the case when the input noise is T_0. Questions and ambiguities arise immediately, however. Is T_0 input to all the input ports, and if so, is the input noise to the different ports correlated? How are the other output ports terminated; is T_0 input to them as well? For differential amplifiers, is the noise input to the physical ports 1 and 2 or to the differential and common modes?

The most significant complicating factor is the correlation matrix of the incident noise waves. In actual use, the input to the amplifier will come from other parts of the circuit, and the noise incident on different ports may well be correlated to some degree. Also, each input port may have a different incident noise temperature. Both these complications are contained in the incident noise correlation matrix. The output noise powers will be given by the appropriate elements of the noise matrix, which depend on the incident noise waves $\left\langle a_i a_j^* \right\rangle$ and therefore on $\boldsymbol{A}$. There are two possible strategies for dealing with the correlations among incident noise waves. The noise figure can be defined to be a function of the incident noise correlation matrix, just as it is a function of the reflection coefficients of the terminations; or a reference value (e.g. the identity matrix) can be chosen for the incident noise correlation matrix, much as we choose a reference noise temperature T_0. However, unlike the case with the reference noise temperature T_0, it would not be possible to compute the noise figure for some other value of $\boldsymbol{A}$ from the noise figure with the reference value $\boldsymbol{A_0}$. Consequently, we choose to treat the noise figure as a function of $\boldsymbol{A}$, the (complex) correlation matrix of the input noise waves.

For the terminations of the output ports, we follow the spirit of the two-port definitions, namely that the noise figure measures the noise added by the amplifier for a given choice of reflection coefficients for the input terminations, but it should not include noise contributions from the various output loads. We (tentatively) adopt the convention that no noise source is connected to the output ports. In practice, it should make little difference since the isolation between the different output ports should be great enough that the output of a given port would be insensitive to whether T_0 is applied to some other output port, especially considering that T_0 is applied to the input channels, which are being amplified.

The definition of the noise figure for a given output channel is then complete. In terms of the notation introduced in Section 8.2, it takes the form

$$F_i(\boldsymbol{\Gamma}, \boldsymbol{A}) = 1 + \frac{1}{k_B T_0} \frac{\{[\mathbf{1} - \boldsymbol{S\Gamma}]^{-1} \hat{N} [\mathbf{1} - \boldsymbol{S\Gamma}]^{-1\dagger}\}_{ii}}{\{[\mathbf{1} - \boldsymbol{S\Gamma}]^{-1} \boldsymbol{SAS}^{\dagger} [\mathbf{1} - \boldsymbol{S\Gamma}]^{-1\dagger}\}_{ii}} \tag{8.12}$$

where the subscript i's indicate the element of the matrix within the braces. This definition reduces to the usual two-port definition, and it embodies the intuitive idea that the noise figure measures how much noise the amplifier adds to a 290-K reference signal. If it seems somewhat formal at this point, it should become clearer in the following sections, when we work through two simple examples in detail.

In addition to defining the noise figure, Eq. (8.12) also constitutes a parameterization of its dependence on the reflection coefficients of the sources and loads, on the incident noise correlation matrix, and on the intrinsic properties of the M-port device itself. The noise parameters of the amplifier are the M^2 independent elements of the intrinsic noise matrix. (The matrix is complex, but it is Hermitian.) For a three-port amplifier, there would be nine real parameters. In the notation of Eqs. (8.8) and (8.9), they would be the three characteristic noise temperatures and three complex correlation functions. In principle, one could develop a multiport parameterization analogous to the IEEE parameterization for the two-port noise figure or effective input noise temperature. That would also require nine real parameters, and that set of parameters could be expressed in terms of the elements of the noise matrix and the S-parameters of the amplifier. That exercise is left for another day (and another author).

8.4 Degradation of Signal-to-Noise Ratio

For a two-port amplifier, the noise figure is a direct measure of the degradation of the signal-to-noise (s/n) ratio. In general, the noise figure of Eq. (8.12) is not the ratio of input s/n to output s/n, but it is not difficult to obtain the expression for that ratio. Let the wave amplitude of the input signal be a_{sig}, and assume that the signal channel is port 1. The input signal power density is $s_{in} = |a_{sig}|^2$, and the output signal power density in port i is given by

$$s_{out} = \{[\mathbf{1} - \boldsymbol{S\Gamma}]^{-1}\boldsymbol{A}_{sig}[\mathbf{1} - \boldsymbol{S\Gamma}]^{-1\dagger}\}_{ii}s_{in}$$

$$\boldsymbol{A}_{sig} = \begin{pmatrix} 1 & 0 & 0 & \dots \\ 0 & 0 & 0 & \dots \\ 0 & 0 & 0 & \dots \\ . & . & . & \dots \end{pmatrix} \tag{8.13}$$

The input noise power density is taken to be k_B times the reference temperature, $n_{in} = k_B T_0$, and the output noise power density in port i is

$$n_{out} = \{k_B T_0[\mathbf{1} - \boldsymbol{S\Gamma}]^{-1}\boldsymbol{SAS}^\dagger[\mathbf{1} - \boldsymbol{S\Gamma}]^{-1\dagger} + [\mathbf{1} - \boldsymbol{S\Gamma}]^{-1}\widehat{\boldsymbol{N}}[\mathbf{1} - \boldsymbol{S\Gamma}]^{-1\dagger}\}_{ii} \tag{8.14}$$

where $\boldsymbol{A}$ is the incident noise correlation matrix for the actual configuration, except that $A_{11} = 1$. The degradation of the signal-to-noise ratio for output port i, which shall be denoted $F_i^{s/n}$, is then given by

$$F_i^{\frac{s}{n}}(\boldsymbol{\Gamma}, \boldsymbol{A}) \equiv \frac{\left(\frac{s}{n}\right)_{in}}{\left(\frac{s}{n}\right)_{out}} = \frac{\{[\mathbf{1} - \boldsymbol{S\Gamma}]^{-1}[k_B T_0 \boldsymbol{SAS}^\dagger + \widehat{\boldsymbol{N}}][\mathbf{1} - \boldsymbol{S\Gamma}]^{-1\dagger}\}_{ii}}{k_B T_0\{[\mathbf{1} - \boldsymbol{S\Gamma}]^{-1}\boldsymbol{SA}_{sig}\boldsymbol{S}^\dagger[\mathbf{1} - \boldsymbol{S\Gamma}]^{-1\dagger}\}_{ii}} \tag{8.15}$$

The difference between this form and Eq. (8.12) is that (8.12) has the full $\boldsymbol{A}$ in the denominator, whereas the $F^{s/n}$ of Eq. (8.15) has only $\boldsymbol{A}_{sig}$. The noise figure of

Eq. (8.12) takes the total noise out (in port i) and divides it by the noise out due to all the incident noise, whereas $F^{s/n}$ divides by the noise out due to the incident noise in the signal channel only. For the $\boldsymbol{\Gamma} = 0$ case, Eq. (8.15) reduces to

$$F_i^{\frac{s}{n}}(\boldsymbol{\Gamma} = \mathbf{0}, \boldsymbol{A}) = \frac{\{[k_B T_0 \boldsymbol{SAS}^\dagger + \widehat{\boldsymbol{N}}]\}_{ii}}{k_B T_0 |S_{i1}|^2} \tag{8.16}$$

and $|S_{i1}|^2$ is the gain from the incident signal channel (1) to the output channel i.

8.5 Three-Port Example – Differential Amplifier with Reflectionless Terminations

8.5.1 Motivation

The developments thus far in this chapter have been rather formal and are sufficiently abstract as to benefit from specific explicit examples. In order to completely characterize the noise properties of a multiport amplifier, to determine its noise figures for general terminations and input noise correlations, it is necessary to determine the complete intrinsic noise matrix. There are, however, specific configurations or choices of terminations that are of interest in their own right. In particular, the case in which all ports have reflectionless terminations is often a useful approximation to the exact actual configuration. Also, it is often useful to have a single number, or a figure of merit, that summarizes or represents an amplifier's properties. The noise figure for a reflectionless input termination is often used for this purpose in the two-port case, and we expect that the noise figure with reflectionless terminations and uncorrelated incident noise can serve a similar purpose for multiport amplifiers. Consequently, the $\boldsymbol{\Gamma} = 0$ examples considered in this and the subsequent section may be of some practical use, as well as clarifying the multiport noise-figure definition.

8.5.2 Characteristic Noise Temperature, Gains, and Effective Input Noise Temperature

A differential amplifier is a three-port device with a single output port whose signal (ideally) is proportional to the difference between the signals at the two input ports. Let port 3 be the output port. Then define input waves $a_\pm$ and S parameters $S_{3\pm}$ to describe the common (+) and differential (−) modes

$$a_\pm \equiv (a_1 \pm a_2)/\sqrt{2}$$
$$S_{3\pm} = (S_{31} \pm S_{32})/\sqrt{2} \tag{8.17}$$

The output amplitude at port 3 is then

$$b_3 = S_{3-}a_- + S_{3+}a_+ + c_3 \tag{8.18}$$

where ideally $S_{3+} = 0$. One immediate, important consequence of the definitions of Eq. (8.17) is that if the noise waves represented by a_1 and a_2 are uncorrelated, then the noise temperatures input to the common and differential modes are equal. Therefore, to obtain different input noise temperatures for the common and differential modes requires correlated noise sources for ports 1 and 2.

We consider the simple case of all ports terminated with reflectionless loads or sources. Since there are no reflections from the terminations, the off-diagonal elements of the noise matrix do not contribute to the output noise at port 3, nor do the characteristic noise temperatures of the input ports, $\widehat{T}_1$ and $\widehat{T}_2$. Only $\widehat{T}_3$, the characteristic noise temperature of port 3, contributes to the output noise, just as in the case of a two port with reflectionless terminations.

If two uncorrelated noise sources with noise temperatures T_1^{in} and T_2^{in} are input to ports 1 and 2, we can use Eq. (8.18) to write the average noise power per unit bandwidth emerging from port 3 as

$$n_3 = k_B\left[G_{31}T_1^{in} + G_{32}T_2^{in} + \widehat{T}_3\right] \tag{8.19}$$

where $G_{31} = |S_{31}|^2$, $G_{32} = |S_{32}|^2$.

The unknown parameters in Eq. (8.19) can be determined from a series of hot–cold measurements similar to the two-port case, an extension of the Y-factor method to the three-port case. Let T_{h1} denote the noise temperature of the hot source connected to port 1, etc. In principle, T_{h1} and T_{h2} could be equal, and T_{c1} and T_{c2} probably will be equal to the ambient temperature and therefore to each other, but we begin with the general case. There are then four different measurements that can be performed. Let $n_{3,hc}$ be the output noise power per unit frequency measured at port 3, for a hot source on port 1 and a cold source on port 2. Define $n_{3,hh}$, $n_{3,ch}$, and $n_{3,cc}$ in a similar manner. The results of the four measurements are then given by

$$\begin{aligned} n_{3,hh} &= k_B[G_{31}T_{h1} + G_{32}T_{h2} + \widehat{T}_3] \quad & n_{3,hc} &= k_B[G_{31}T_{h1} + G_{32}T_{c2} + \widehat{T}_3] \\ n_{3,ch} &= k_B[G_{31}T_{c1} + G_{32}T_{h2} + \widehat{T}_3] \quad & n_{3,cc} &= k_B[G_{31}T_{c1} + G_{32}T_{c2} + \widehat{T}_3] \end{aligned} \tag{8.20}$$

There are only three unknowns in Eq. (8.20), i.e. G_{31}, G_{32}, and $\widehat{T}_3$; and consequently, the equations are not all independent. Indeed, one notes that

$$n_{3hh} + n_{3cc} = n_{3hc} + n_{3ch} \tag{8.21}$$

Therefore, it is sufficient to measure only three of the four hot–cold combinations to determine the gains and $\widehat{T}_3$. (Although in principle it is sufficient to measure only three, in practice it may be preferable to measure all four combinations and fit for a best solution.) The set of *hc*, *ch*, and *cc* may give slightly better accuracy, and it requires only one hot noise source; so, we shall work with that set. The measured values for the gains are then given by

$$G_{31} = \frac{n_{3,hc} - n_{3,cc}}{k_B(T_{h1} - T_{c1})}$$
$$G_{31} = \frac{n_{3,ch} - n_{3,cc}}{k_B(T_{h2} - T_{c2})} \tag{8.22}$$

and the intrinsic output noise temperature for port 3 is given by

$$k_B\hat{T}_3 = \frac{(T_{h1}T_{h2} - T_{c1}T_{c2})}{(T_{h1} - T_{c1})(T_{h2} - T_{c2})}n_{3,cc} - \frac{T_{c1}}{(T_{h1} - T_{c1})}n_{3,hc} - \frac{T_{c2}}{(T_{h2} - T_{c2})}n_{3,ch} \tag{8.23}$$

The equivalent input temperature, which is equal for the two input ports, is given by

$$T_e = \frac{\hat{T}_3}{G_{31} + G_{32}} \tag{8.24}$$

with $\hat{T}_3$ given by Eq. (8.23). If we assume that the two cold temperatures are equal and equal to T_c, T_e can be written as

$$T_e = \frac{T_{h1}T_{h2} - (T_{h1}Y_{ch} - T_{h2}Y_{hc})T_c + Y_{hh}T_c^2}{(Y_{ch} - 1)T_{h1} + (Y_{hc} - 1)T_{h2} - (Y_{hh} - 1)T_c} \tag{8.25}$$

where $Y_{hc} = n_{3,hc}/n_{3,cc}$, etc. Although $n_{3,hh}$ may not have been measured, Y_{hh} can be determined from Eq. (8.21), $Y_{hh} = Y_{hc} + Y_{ch} - 1$. If we further assume that only one hot source is used, so that $T_{h1} = T_{h2} = T_h$, then Eq. (8.25) reduces to

$$T_e = \frac{T_h - Y_{hh}T_c}{Y_{hh} - 1} \tag{8.26}$$

which is the reassuringly familiar two-port result of Section 4.3, with Y_{hh} playing the role of the two-port Y.

Equation (8.26) indicates that T_e can be determined either from the set of three measurements (*hc*, *ch*, and *cc*) or from just two measurements (*hh* and *cc*) if $T_{h1} = T_{h2}$. If only *hh* and *cc* are measured, we can still determine the sum of the gains and $\hat{T}_3$

$$G_{31} + G_{32} = \frac{n_{hh} - n_{cc}}{k_B(T_h - T_c)}$$
$$\hat{T}_3 = \frac{k_B(T_h - Y_{hh}T_c)(T_h - T_c)}{(Y_{hh} - 1)(n_{hh} - n_{cc})} \tag{8.27}$$

but we cannot determine either gain separately as in Eq. (8.22).

The discussion in this section has not yet treated the differential or common mode. From $G_{3\pm} = |S_{3\pm}|^2$ and Eqs. (8.17) and (8.18), it follows that

$$G_{3+} + G_{3-} = G_{31} + G_{32} \tag{8.28}$$

Since $\hat{T}_3$ is the same no matter how we describe the input ports and since the sum of the gains is the same, the effective input noise temperature in the differential and common modes is the same as for ports 1 and 2. The hot-cold measurements with uncorrelated sources, described above, are therefore sufficient to determine $\hat{T}_3$, T_e, and $G_{3+} + G_{3-}$ for the differential and common modes, but not G_{3+} or G_{3-} individually. Since G_{3-} is designed to be much larger than G_{3+}, we might use the approximation $G_{3-} \approx G_{31} + G_{32}$, but it would be useful to measure G_{3-} or G_{3+} independently. Using noise to measure G_{3-} or G_{3+} requires correlated noise input to ports 1 and 2. If $a_1 = a_2$, then $a_- = 0$ and $a_+ = \sqrt{2}a_1$, which in turn leads to input noise temperatures of $T_- = 0$ and $T_+ = 2T_1$. If the measured noise power density out of port 3 in such a measurement is called $n_{3,+}$, then

$$n_{3,+} = k_B[2G_{3+}T_1 + \hat{T}_3] \tag{8.29}$$

from which it follows that

$$G_{3+} = \frac{(G_{31} + G_{32})}{2T_1}[T_c Y_+ + T_e(Y_+ - 1)] \tag{8.30}$$

where $Y_+ = n_{3,+}/n_{3,cc}$. All quantities on the right-hand side, except Y_+, can be determined from the uncorrelated measurements described above, and therefore measurement of Y_+ determines G_{3+}.

To summarize the three-port matched case, with just one hot source and two equal-temperature cold sources, a set of three measurements (*hc*, *ch*, *cc*) with uncorrelated input noise will determine $\hat{T}_3$, T_e, G_{31}, G_{32}, and $G_{3+} + G_{3-}$. (Obviously, if a second hot source is available, ***hh*** could be done as a consistency check or to reduce the measurement uncertainty.) If two equal-temperature hot sources and two equal-temperature cold sources are available, then just two measurements (***hh*** and *cc*) suffice to determine $\hat{T}_3$, T_e, $G_{31} + G_{32}$, and $G_{3+} + G_{3-}$, but not any individual gain. To determine G_{3-} *or* G_{3+} individually (in a noise measurement) requires the noise input to ports 1 and 2 to be correlated. Of course, as in the two-port case, the gains could instead be determined by vector network analyzer (VNA) measurements.

8.5.3 Noise Figure

Once all the relevant parameters have been measured, as in the preceding subsection, we can compute the noise figure of the differential amplifier. For two input ports and one output port, all with reflectionless terminations, and no correlation between the incident noise waves ($\boldsymbol{A} = \mathbf{1}$), Eq. (8.12) reduces to

$$F_3 = 1 + \frac{\hat{T}_3}{(G_{31} + G_{32})T_0} = 1 + \frac{T_e}{T_0} \tag{8.31}$$

Note that this noise factor does not require separate measurement of G_{3-} or G_{3+} and thus does not require any measurements with correlated noise input.

Equation (8.31) gives the ratio of the total output noise to that portion of the output noise that is due to all input noise for the particular case considered, $\boldsymbol{\Gamma} = 0$, $\boldsymbol{A} = \mathbf{1}$. As discussed in Section 8.4, however, it does not give the degradation of the s/n ratio. That is given by Eq. (8.15), or Eq. (8.16) for $\boldsymbol{\Gamma} = 0$. If the input channel is the differential mode, Eq. (8.16) takes the form

$$F^{\frac{s}{n}}(\boldsymbol{\Gamma} = \mathbf{0}, \boldsymbol{A} = \mathbf{1}) = \frac{(G_{3-} + G_{3+})T_0 + \widehat{T}_3}{G_{3-}T_0} = \left(1 + \frac{G_{3+}}{G_{3-}}\right)\left(1 + \frac{T_e}{T_0}\right) \tag{8.32}$$

This differs from the noise figure of Eq. (8.31) by the factor of $(1 + G_{3+}/G_{3-})$, which renders $F^{s/n}$ more difficult to measure (unless one is willing and able to make the approximation $(1 + G_{3+}/G_{3-}) \approx 1$). It does, however, provide a better measure of the amplifier's s/n performance. The gains G_{3-} and G_{3+} can be determined from noise measurements with correlated noise incident on the two input ports or from VNA measurements.

It is also interesting to consider $F^{s/n}$ for the case of $\boldsymbol{A} \neq \mathbf{1}$, which would be appropriate for a circuit configuration in which the noise incident on the different ports of the differential amplifier was correlated. In this case, Eq. (8.16) takes the form

$$F^{\frac{s}{n}}(\boldsymbol{\Gamma} = \mathbf{0}, \boldsymbol{A}) = 1 + \frac{\widehat{T}_3 + T_0\left[G_{3+}A_{++} + 2Re\left(S_{3+}S_{3-}^{*}A_{\pm}\right)\right]}{T_0 G_{3-}} \tag{8.33}$$

The only noise parameter of the amplifier that enters Eq. (8.33) is $\widehat{T}_3$. The other parameters needed are the elements of the correlation matrix of the incident noise, coming from other parts of the circuit in which the amplifier is embedded, and the scattering parameters of the amplifier, not just the gains. The incident noise correlation matrix element A_{11} is taken equal to one, as prescribed by the convention for defining the s/n noise factor.

8.5.4 Practical Applications

The traditional method for measuring the noise performance of a differential amplifier employs a balun, thereby reducing the measurement problem to a single output port [13]. More recently, new methods based on the mixed-mode formulation [3] or on the wave approach [6] have been developed and implemented [13–16]. Central to the measurements using the wave approach is the ability to measure correlations between the different noise waves. As new instrumentation with lower levels of phase noise [17] becomes available, we may expect such measurements to become more common – and more accurate.

8.6 Four-Port Example with Reflectionless Terminations

As a further example of the formalism, we consider an amplifier with two input and two output ports (or modes), such as the mixed-mode two-port treated in [3].

To make the example concrete, we take port 3 to be the differential output mode and port 4 to be the common output mode. Ports 1 and 2 are taken to correspond to two physically separate input ports, though they could just as well be the differential and common input modes. Again, we treat only the reflectionless case with uncorrelated incident noise, $\boldsymbol{\Gamma} = \mathbf{0}$, $\boldsymbol{A}$ diagonal. Equation (8.12) then reduces to

$$F_3 = 1 + \frac{\langle |c_3|^2 \rangle}{|S_{31}|^2 \langle |\hat{a}_1|^2 \rangle + |S_{32}|^2 \langle |\hat{a}_2|^2 \rangle} = 1 + \frac{\hat{T}_3}{(G_{31} + G_{32})T_0} \tag{8.34}$$

for the noise figure of port 3. Port 4 has its own noise figure, given by a similar equation, but with $3 \to 4$. Determination of F_3 then requires determination of $\hat{T}_3$, the characteristic noise temperature for port 3, as well as the sum of the two gains, $(G_{31} + G_{32})$ or $(G_{3-} + G_{3+})$.

The characteristic noise temperatures and sums of gains can be determined by a series of hot-cold measurements, as in the preceding section. If we first measure the output noise in port 3 while connecting hot and cold loads to the input ports, we again obtain the set of equations in Eqs. (8.20) and (8.21). Solving for the gains again yields Eq. (8.22), and solving for $\hat{T}_3$ again yields Eq. (8.23). Similarly, if we measure the output noise in port 4 while connecting hot and cold loads to the input ports, we obtain the same equations for the gains and for $\hat{T}_4$, with 3 replaced by 4. In the measurements for output port 3 or 4, the other port is terminated in a reflectionless load.

Thus far, everything is essentially the same as in the three-port case. The gains and the intrinsic output noise temperatures $\hat{T}_3$ and $\hat{T}_4$ are determined by a series of hot and cold measurements. However, a nuance arises when we attempt to compute the effective input temperature. When there are two output ports, the two input ports in general do not have the same effective input temperature. The equations that define T_{1e} and T_{2e} are

$$\begin{aligned} \hat{T}_3 &= G_{31}T_{1e} + G_{32}T_{2e} \\ \hat{T}_4 &= G_{41}T_{1e} + G_{42}T_{2e} \end{aligned} \tag{8.35}$$

Solving for the effective input noise temperatures, we obtain

$$\begin{aligned} T_{1e} &= \frac{G_{42}\hat{T}_3 - G_{32}\hat{T}_4}{\Delta_G} \\ T_{2e} &= \frac{-G_{41}\hat{T}_3 + G_{31}\hat{T}_4}{\Delta_G} \\ \Delta_G &= G_{31}G_{42} - G_{32}G_{41} \end{aligned} \tag{8.36}$$

For the s/n degradation, the expressions are similar to the three-port case,

$$F_3^{s/n}(\boldsymbol{\Gamma} = \mathbf{0}, \boldsymbol{A}) = 1 + \frac{\hat{T}_3 + T_0\left[G_{3+}A_{++} + 2Re\left(S_{3+}S_{3-}^*A_{+-}\right)\right]}{T_0G_{3-}}$$

$$F_3^{\frac{s}{n}}(\boldsymbol{\Gamma} = \mathbf{0}, \boldsymbol{A} = \mathbf{1}) = \frac{(G_{3-} + G_{3+})T_0 + \hat{T}_3}{G_{3-}T_0} \tag{8.37}$$

The equations for port 4 are obtained by replacing 3 by 4 in Eq. (8.37).

In summary, the four-port case introduces two complications absent for three ports. The obvious complication is that there are two noise figures, one for each output port. Each noise figure can be measured in a manner similar to the three-port (or two-port) case, with a series of hot-cold measurements. (In fact, the two noise figures could be measured simultaneously.) The second complication, which might not have been expected, is that there must be two different effective input noise temperatures. These are given in Eq. (8.36) in terms of the gains and characteristic noise temperatures of the output ports. The expressions for the signal-to-noise figures are similar to the three-port case.

The definition of the effective input noise temperature becomes even more difficult when there are more output than input ports. In that case, the generalization of Eq. (8.35) will usually not admit a solution for the effective input noise temperatures for the various input ports, the T_{ie}'s, and we would have to define a different effective input noise temperature for each output (and input) port. One might then legitimately question the utility of even introducing effective input noise temperatures.

References

1 H.A. Haus and R. Adler, *Circuit Theory of Linear Noisy Networks*, Chapter 5.4, Wiley, New York, 1959.

2 H.A. Haus, R. Adler, R.S. Engelbrecht et al., "Description of the noise performance of amplifiers and receiving systems," *Proceedings of the IEEE*, **51**, pp. 436–442 (March 1963), 10.1109/PROC.1963.1846.

3 D.E. Bockelman and W.R. Eisenstadt, "Combined differential and common-mode scattering parameters: theory and simulation," *IEEE Transactions on Microwave Theory and Techniques*, **43**, no. 7, pp. 1530–1539 (July 1995).

4 H. Bosma, "On the theory of linear noise systems," *Philips Research Reports Supplements*, **10**, p. 191 (1967).

5 D.F. Wait, "Thermal noise from a passive linear multiport," *IEEE Transactions on Microwave Theory and Techniques*, **16**, no. 9, pp. 687–691 (September 1968).

6 J. Randa, "Noise characterization of multiport amplifiers," *IEEE Transactions on Microwave Theory and Techniques*, **49**, no. 10, pp. 1757–1763 (October 2001).

7 I. Corbella, F. Torres, A. Camps, N. Duffo, M. Vall-llossera, K. Rautiainen, M. Martín-Neira, and A. Colliander, "Analysis of correlation and total power radiometer front-ends using noise waves, *IEEE Transactions on Geoscience and Remote Sensing*, **43**, no. 11, pp. 2459–2452 (November 2005). doi: 10.1109/TGRS.2005.847912.

8 R. Meys, "A wave approach to the noise properties of linear microwave devices," *IEEE Transactions on Microwave Theory and Techniques*, **MTT-26**, pp. 34–37 (January 1978).

9 P. Penfield, "Wave representation of amplifier noise," *IRE Transactions on Circuit Theory*, **CT-9**, pp. 84–86 (March 1962).

10 S.W. Wedge and D.B. Rutledge, "Wave techniques for noise modeling and measurement," *IEEE Transactions on Microwave Theory and Techniques*, **40**, no.11, pp. 2004–2012 (November 1992).

11 J. Randa, "Noise temperature measurements on wafer," National Institute of Standard and Technology Technical Note 1390 (March 1997). doi: 10.6028/NIST.TN.1390.

12 R.E. Collin, *Field Theory of Guided Waves*, 2nd edn., IEEE Press, Piscataway, NJ, 1991, pp. 342–349.

13 J. Martens, "Differential noise measurements: sensitivities and uncertainties with direct correlation- and balun-based methods," in *93rd ARFTG Microwave Measurement Conference*, 2019, pp. 1–5. doi: 10.1109/ARFTG.2019.8739166.

14 J. Dunsmore and S. Wood, "Vector corrected noise figure and noise parameter measurements of differential amplifiers," in *2009 European Microwave Conference (EuMC)*, pp. 707–710. doi: 10.23919/EUMC.2009.5296507.

15 Y. Andee, C. Arnaud, F. Graux and F. Danneville, "De-embedding differential noise figure using the correlation of output noise waves," in *85th ARFTG Microwave Measurement Conference*, 2015, pp. 1–4. doi: 10.1109/ARFTG.2015.7162917.

16 L. Boglione, "Generalized determination of device noise parameters," *IEEE Transactions on Microwave Theory and Techniques*, **65**, no. 10, pp. 4014–4025 (October 2017). doi: 10.1109/TMTT.2017.2709320.

17 W.F. Walls, "Cross-correlation phase noise measurements," in *Proceedings of the 1992 IEEE Frequency Control Symposium*, 1992, pp. 257–261. doi: 10.1109/FREQ.1992.270007.

9

Remote Sensing Connection

9.1 Introduction

Microwave radiometry traces its origins to the early days of radio astronomy, and so closing this book with a chapter on remote sensing also closes the circle back to the roots of the field. We will not deal with the actual practice of remote sensing; that would be (and has been [1]) a multivolume work in itself. Instead, we will deal with the connection of remote sensing to the laboratory noise measurements that we have discussed in the earlier chapters of this book, and to the support that the laboratory measurements can provide to microwave remote-sensing measurements in the field. When a particular remote-sensing context is assumed, it will be that of microwave remote sensing of the earth from instruments on satellites, but the general results have broader applicability.

An important consideration in microwave remote-sensing measurements is traceability to fundamental standards, already discussed in Chapter 3.6. Such traceability offers a number of benefits, all deriving from the existence of a stable, common point of reference to which different measurements by different instruments at different times can be compared. Thus, for example, measurements from two different satellites could be meaningfully compared and (if necessary) reconciled if both were traceable to fundamental standards. If the standards are derived from basic physical principles, then they need not even be realized at the same laboratory. The remote-sensing instruments could be directly traceable to the realizations of the same fundamental physical quantity at the national measurement institutes (NMIs) of different countries. At least in principle, this would allow the comparison and reconciliation of data from different nations or organizations. It would also provide a stable reference point that would allow the comparison of data from satellites flown years or decades apart, a critical issue for studying long-term phenomena such as climate change.

Without any clarification or specification, traceability can be a weak, almost nebulous, term in the context of remote sensing. It can refer to a single component

Precision Measurement of Microwave Thermal Noise, First Edition. James Randa.

or set of components, such as the temperature sensors in a calibration target, or a voltmeter used to read the output of a detector; or it can refer to an entire instrument being calibrated against NMI standards, as is done at infrared, optical, and ultraviolet frequencies at National Institute of Standards and Technology (NIST) [2]. The two key considerations we should bear in mind in discussing traceability are: (i) What is the physical quantity whose measurement is traceable? (ii) What is the uncertainty associated with the traceability chain? These two questions are linked, of course.

In the case of microwave remote sensing, the physical quantity being measured is typically the received brightness temperature. The fact that the measurement of detector voltages or temperatures within a calibration target are NMI-traceable with some small uncertainty says little about the measured brightness temperature. The intervening steps from temperature within the target or detector voltage to brightness temperature are open to question: they must be verified, and the uncertainties estimated. The more such steps there are, the larger the uncertainty becomes. In addition, each such step and the associated uncertainty between the traceable quantity (temperature within the target or a voltage) is open to question by doubters. The more that can be included within the traceability and the smaller the uncertainties, the better.

The usual way to calibrate a remote-sensing radiometer is to have it view two standard targets whose brightness temperatures are known, similar to the procedure described for coaxial or waveguide radiometers described in Chapter 3. The switching between targets can be effected by using mirrors, or it can be done by moving the entire antenna, pointing it at the different targets sequentially. For example, in a satellite-based radiometer, for one calibration target the satellite can be rolled so that it looks at cold space (This procedure carries an attendant risk that one may not be able to turn the satellite back to view the target of interest.) We shall call this calibration procedure the standard target method. For this method, the traceability is through the target brightness temperature, and uncertainties arise due to possible temperature gradients in the target, uncertainty in the target emissivity, corruption of the received power by radiation from other sources, and power received in antenna side lobes, among others.

An alternative way to achieve traceability for remote sensing measurements is to utilize the standards and radiometers developed for transmission line (waveguide or coaxial) measurements, described earlier in this book, and attach an antenna in place of the device under test (DUT). If the relevant characteristics of the antenna were known, the brightness temperature incident on it could be measured, traceable to the primary noise standards. We will refer to this as the "standard-radiometer" method or approach, as opposed to the standard-target approach. Although it is usually not practical to use a standard radiometer directly in a remote-sensing application, it can be used to measure the brightness temperature of a standard target, either as a calibration of the target or as a check of the target's brightness temperature.

To implement the standard-radiometer approach, we must first determine how to characterize the receiving antenna, and we address that task in Sections 9.2 and 9.3.

9.2 Theory for Standard Radiometer

We wish to convert (reversibly of course) current radiometers, which are used to measure coaxial or waveguide noise sources, into standard remote sensing radiometers that can then be used to link microwave remote-sensing radiometer measurements to primary thermal noise standards. This section sets out the general framework for doing so and indicates the particular antenna properties that must be determined (in some way) in order to link the remote-sensing measurements to the primary standards. We follow the development presented in [3], which in turn follows the treatment in Chapter 4 of Volume 1 of Ulaby et al. [1], except that it does not rely on the Rayleigh–Jeans approximation as they do. Throughout the development, we assume far-field conditions; possible departures from this must be included in the uncertainty analysis.

The basic configuration of interest is shown in Figure 9.1. It is similar to the usual configuration for measuring the noise temperature of a noise source except that an antenna replaces the noise source. The antenna is placed in a shielded, temperature-controlled anechoic chamber, to control the background radiation that may be received. The output noise temperature of the antenna (at plane x) can

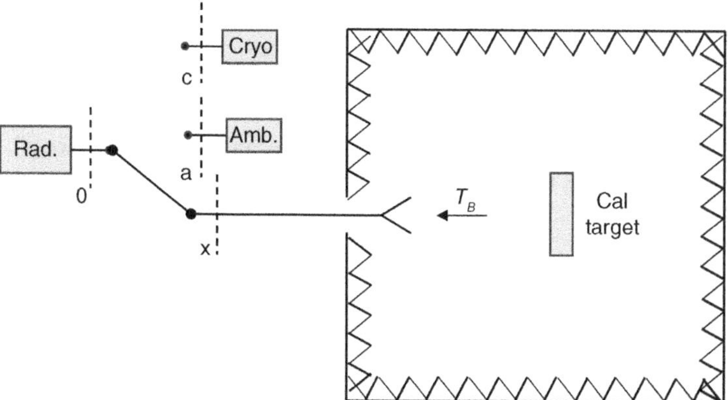

Figure 9.1 Standard-radiometer configuration. Source: From Randa et al. [4]/U.S. Department of Commerce/ Public Domian.

be measured in the same way as would the noise temperature of a noise source. Our task is to relate the noise temperature measured at plane x to the radiation intensity (or brightness or brightness temperature [5]) incident on the antenna. The brightness $B(\theta, \varphi)$ is the power per unit area and solid angle incident on (or emitted from) a surface, and the spectral brightness $B_f(\theta, \varphi)$ is the brightness per unit frequency. In terms of the spectral brightness, the differential power received by an antenna from differential solid angle $d\Omega$ in differential frequency interval df can be written as

$$dp = A_{eff}B_f(\theta, \phi)F_n(\theta, \phi)d\Omega df \tag{9.1}$$

where A_{eff} is the effective aperture of the antenna, and $F_n(\theta,\varphi)$ is its normalized antenna pattern. This assumes that the polarization of the incident radiation is matched to the antenna. For unpolarized incident radiation the total power received in the frequency interval Δf is

$$P = \frac{1}{2}A_{eff}\int_f^{f+\Delta f}\int_{4\pi} B_f(\theta, \phi)F_n(\theta, \phi)d\Omega df \tag{9.2}$$

where the ½ arises from the average over incident polarization [1]. In principle, of course, F_n is also a function of frequency, but we assume that Δf is small enough that F_n is effectively constant across the range of the frequency integration. From Eq. (9.2) we can also identify the spectral power P_f received by the antenna,

$$P_f = \frac{1}{2}A_{eff}\int_{4\pi} B_f(\theta, \phi)F_n(\theta, \phi)d\Omega \tag{9.3}$$

To introduce the concept of brightness temperature, we consider the case of an antenna receiving radiation from an ideal black body. The spectral brightness density emitted by an ideal black body at temperature T is uniform in all directions and is given by

$$B_f = \frac{2}{\lambda^2}\left(\frac{hf}{e^{\frac{hf}{kT}} - 1}\right) \tag{9.4}$$

where h and k are Planck's and Boltzmann's constants, and λ is the wavelength. For $hf/kT \ll 1$, this assumes the familiar Rayleigh–Jeans form

$$B_f \approx \frac{2kT}{\lambda^2} \tag{9.5}$$

where T is the physical temperature of the ideal black body.

Because of the approximate linear relationship between B_f and T for a black body, and because the physical temperature of the target is typically the quantity of interest in microwave remote sensing, it is common to define and work with a "brightness temperature" T_B rather than the brightness itself. The definition most commonly used in microwave remote sensing (but which we will *not* use) is what can be called the equivalent blackbody definition, i.e. that the brightness temperature is the temperature of an ideal black body that would give rise to the observed brightness. Thus, Eq. (9.4) is used for any radiating body (not just an ideal black body), with T defining the brightness temperature. With this definition, the brightness temperature of a black body is equal to its physical temperature, but the relationship between B_f and T_B expressed by Eq. (9.5) is valid only in the Rayleigh–Jeans approximation. This is rather inconvenient, as will become evident below. The alternative [4] is to use Eq. (9.5) to *define* T_B,

$$T_B(\theta, \phi) \equiv \frac{\lambda^2 B_f(\theta, \phi)}{2k} \tag{9.6}$$

which we will call the power definition. This is essentially the same discussion and the same choice as was made with *noise* temperature in Chapter 1. With the power definition, the correspondence between the brightness temperature and the physical temperature for an ideal black body is only approximate, but this definition is much more convenient for use in common equations. For this reason, and to be consistent with the definition of noise temperature, we adopt the power definition of T_B, Eq. (9.6). The relation between T_B and the physical temperature of an ideal black body is obtained by using Eq. (9.4) for B_f in Eq. (9.6). For a non-black body, with emissivity less than one, a factor of the emissivity ε is inserted into the right side of Eq. (9.4), and Eq. (9.6) is unchanged. Note that for non-black bodies, the emissivity in general is a function of the incidence angles.

Using Eq. (9.6) in Eq. (9.2) and assuming that Δf is small, we have

$$P = \frac{kA_{eff}\Delta f}{\lambda^2} \int_{4\pi} T_B(\theta, \phi) F_n(\theta, \phi) d\Omega \tag{9.7}$$

Had we adopted the equivalent black body definition of the noise temperature, Eq. (9.7) would be complicated by a Planck factor. Similarly, all the equations below in which temperatures are added – or integrated – depend on the power definition of the brightness temperature. (Powers are additive; equivalent black-body temperatures are not.) If we define the input antenna temperature $T_{A,in}$ by

$$T_{A,in} \equiv \frac{A_{eff}}{\lambda^2} \int_{4\pi} T_B(\theta, \phi) F_n(\theta, \phi) d\Omega \tag{9.8}$$

then Eq. (9.7) takes the familiar form for the available power from a passive impedance at noise temperature $T_{A,in}$,

$$P = kT_{A,in}\Delta f \tag{9.9}$$

Thus the input antenna temperature $T_{A,in}$ is the noise temperature (defined as the available noise spectral power divided by Boltzmann's constant) at the input (aperture) plane of the antenna. Equations (9.8) and (9.9) are the sort of relationship that we need. They give the noise power delivered to the radiometer by the antenna in terms of the incident radiation and the antenna properties. Some work still remains to be done, however. Eq. (9.8) for the input antenna temperature can be put in a more useful form by using the relationship between the (lossless) effective area and the maximum directivity D_0, $A_{eff} = \lambda^2 D_0/4\pi$. The maximum directivity is, in turn, related to the pattern solid angle Ω_p by $D_0 = 4\pi/\Omega_p$, where

$$\Omega_p = \int_{4\pi} F_n(\theta, \phi) d\Omega \tag{9.10}$$

The effective area can therefore be written as

$$A_{eff} = \frac{\lambda^2}{\Omega_p} \tag{9.11}$$

and the input antenna temperature assumes the form

$$T_{A,in} = \frac{\int_{4\pi} T_B(\theta, \phi) F_n(\theta, \phi) d\Omega}{\Omega_p} \tag{9.12}$$

Equation (9.12) represents the noise temperature at the antenna input plane, whereas the radiometer measures the noise temperature at the output of the antenna, plane x in Figure 9.1. The two are related by Eq. (1.28),

$$T_{A,out} = \alpha T_{A,in} + (1 - \alpha) T_a \tag{9.13}$$

where α is the available power ratio between the antenna aperture and the output plane (plane x in Figure 9.1), and T_a is the noise temperature corresponding to the physical temperature of the antenna, which is assumed to be ambient temperature. Reference [1] (and most of the remote-sensing community, seemingly) does not worry about subtleties such as the distinction between available power ratio and efficiency (delivered power ratio). They use the efficiency η_l rather than the available power ratio α in Eq. (9.13). They then identify η_l as the inverse of the "loss factor" L, $\eta_l = 1/L$. From our experience with using adapters in noise measurements, this is a reasonable approximation, assuming low loss and good matching. Some care will be required in computing or measuring the loss, to ensure that the correct quantity is obtained, and differences between α and the quantity measured or calculated will need to be accounted for in the uncertainty analysis.

The input antenna temperature of Eqs. (9.12) and (9.13) integrates over the full 4π solid angle. It therefore includes contributions from both the main beam and from any side lobes, as well as from the target of interest and any background radiation. In common remote-sensing applications, the solid-angle integration is separated into an integration over the main lobe plus an integration over everything else. Thus the main beam *defines* the target of interest. In our case, we will be viewing a calibration target, which may intercept only a part of the main beam, or it may intercept more of the pattern than just the main lobe. We wish to separate the calibration-target component of the brightness temperature from everything else. We therefore separate $T_{A,in}$ into two parts, one containing the integration over the solid angle subtended by the target of interest, and the other containing the rest of the 4π solid angle. We do this for both the input antenna temperature and for the pattern solid angle, Eq. (9.10), and we define a target contribution T_T and a background contribution T_{BG} to $T_{A,in}$,

$$\begin{aligned} T_{A,in} &= T_T + T_{BG} \\ T_T &= \frac{\int_{target} T_B(\theta,\varphi)F_n(\theta,\varphi)d\Omega}{\Omega_p} \\ T_{BG} &= \frac{\int_{other} T_B(\theta,\varphi)F_n(\theta,\varphi)d\Omega}{\Omega_p} \end{aligned} \tag{9.14}$$

If we further define an "illumination efficiency" η_{IE},

$$\eta_{IE} \equiv \frac{\int_{target} F_n(\theta,\varphi)d\Omega}{\Omega_p} \tag{9.15}$$

we can write

$$T_{A,in} = \eta_{IE}\overline{T_T} + (1-\eta_{IE})\overline{T_{BG}} \tag{9.16}$$

where

$$\overline{T_T} = \frac{\int_{target} T_B(\theta,\varphi)F_n(\theta,\varphi)d\Omega}{\int_{target} F_n(\theta,\varphi)d\Omega} \tag{9.17}$$

$$\overline{T_{BG}} = \frac{\int_{other} T_B(\theta,\varphi)F_n(\theta,\varphi)d\Omega}{\int_{other} F_n(\theta,\phi)d\Omega}$$

In Eq. (9.17), $\overline{T_T}$ is the incident brightness temperature averaged over the portion of the antenna pattern corresponding to the target solid angle, which is the quantity of interest for a standard radiometer. The illumination efficiency η_{IE} is the fraction of the full antenna pattern that is subtended by the target of interest. The quantity $\overline{T_{BG}}$ is the average brightness temperature from directions other than the target.

We can use Eq. (9.16) in Eq. (9.17) to obtain an equation for $\overline{T_T}$ in terms of measured quantities,

$$\overline{T_T} = \frac{1}{\alpha\eta_{IE}}T_{A,out} - \frac{(1-\eta_{IE})}{\eta_{IE}}\overline{T_{BG}} - \frac{(1-\alpha)}{\alpha\eta_{IE}}T_a \tag{9.18}$$

To control the effect of $\overline{T_{BG}}$ in Eq. (9.19), one needs to control the environment in which the standard radiometer operates. This can be done using an enclosure with absorptive walls, maintained at room temperature, which will also be the temperature of the antenna. In that case $\overline{T_{BG}} = T_a$, and Eq. (9.18) becomes

$$\overline{T_T} = T_a + \frac{1}{\alpha\eta_{IE}}(T_{A,out} - T_a) \tag{9.19}$$

Equation (9.19) is the desired result for standard-radiometer measurements. It allows one to determine $\overline{T_T}$, the average incident brightness temperature received from the target, in terms of $T_{A,out}$, α, η_{IE}, and T_a. In Eq. (9.19), $T_{A,out}$ is the noise temperature at the output of the antenna, measured by the radiometer; α is the available power ratio between the antenna aperture and its output, equal approximately to $1/L$; η_{IE} is the illumination efficiency, defined in Eq. (9.15), and determined from the normalized antenna pattern; and T_a is the noise temperature corresponding to the physical temperature of the antenna and the enclosure.

9.3 Standard-Radiometer Measurements

9.3.1 Determination of α

In order to use Eq. (9.19) to measure target brightness temperature, it is necessary to measure or estimate α and η_{IE}. For α, the approximation $1/L$ can be used, with any small departure from this subsumed into the uncertainty. Assuming one uses a simple antenna, such as a standard-gain horn, ohmic losses are best determined by calculation, using any of a number of software packages. For common, commercial standard-gain horns the ohmic losses are less than 0.03 dB for the worst-case conductivity. The major uncertainty contribution is due to the uncertainty in the electrical conductivity, and it leads to a fractional uncertainty of about 0.5% in α.

9.3.2 Determination of Illumination Efficiency, η_{IE}

Two methods have been used to determine η_{IE}. The first is to measure the full antenna pattern and to evaluate the integral in Eq. (9.15) numerically. This was done in [4], measuring a standard-gain horn on the NIST spherical probe pattern range. The forward pattern was measured in 2° steps in θ and 5° steps in ϕ. It was then integrated over the angle subtended by the target for different antenna-target separation distances. At the closest distance calculated, 50 cm, the illumination efficiency was 0.98, assuming no near-field effects, and it decreased with separation distance.

A second method to determine the illumination efficiency was developed in [6]. It rewrites Eq. (9.18) as

$$T_{A,out} = \alpha\eta_{IE}\overline{T_T} + [(1-\eta_{IE})\alpha\overline{T_{BG}} + (1-\alpha)T_a] \tag{9.20}$$

and then notes that if the antenna output $T_{A,out}$ is measured as the target temperature $\overline{T_T}$ is varied, the slope and intercept of the resulting line can be used to determine η_{IE}. There are several subtleties, but they are worked out in [6]. The uncertainty in η_{IE} depended on the separation distance, but it was generally less than 0.009.

The second method achieves comparable uncertainties to the first method, and it includes any near-field effects. It also does not require access to any test range beyond what would be used for the standard-radiometer measurements themselves. It does not provide any additional information about the test antenna, but that is not needed for the standard-radiometer application.

9.3.2.1 Measurements of a Standard Target

In developing and demonstrating the second method for determining η_{IE}, reference [6] also measured the brightness temperature of a standard target (independent of the determination of η_{IE} [7]). A WR-42 standard-gain horn was used as the antenna, and a borrowed circular blackbody calibration target was used as the target. Very good agreement was found between the measured brightness temperature and its nominal value, particularly at small separation distances. The uncertainty in the brightness-temperature measurements was 1 K at small separation distance. The results and uncertainties indicate that the standard radiometer approach can be valuable in calibrating or checking the brightness temperature of standard blackbody targets in a laboratory environment and providing a traceability chain to fundamental noise standards. It has not yet been tested or implemented in a thermal-vacuum chamber that reproduces conditions encountered by a space-borne radiometer and calibration target.

9.4 Standard-Target Design

While the standard-radiometer method can be used to check or calibrate a calibration target, which is then used to calibrate a remote-sensing radiometer, by far the more common approach is to design and construct a target from first principles, and to then use this target as a primary standard in radiometer calibration. Many such calibration targets have been built, including at NIST [8, 9].

The targets typically consist of a high-emissivity surface material mounted on a metal substrate whose high thermal conductivity distributes the heat and minimizes (lateral) thermal gradients. The metal backing is heated (or cooled) from behind. The geometry of the surface is designed to further increase the absorptivity (and hence the emissivity), through the use of wedge, pyramidal, or cone structures. The geometrical design parameters are usually chosen by computer modeling of the reflectivity of the surface for various sizes and shapes of structures for the intended frequencies of operation. The electromagnetic properties of possible coating materials are measured and compared in order to minimize the reflectivity. Temperature sensors, typically platinum resistance thermometers (PRTs), are embedded in the substrate from behind to monitor the target temperature. The PRTs should be distributed laterally to check for temperature gradients and should be embedded close to the surface in order to most nearly measure the surface temperature, rather than the temperature at a point in the bulk substrate material.

The geometry of the target often consists of pyramids or wedges mounted on a flat substrate, similar to a section of the wall of an anechoic chamber. The targets designed at NIST are conical cavities, with the inside of the cone used as the target. The cone is coated on the inside with a layered absorptive material and heated (or cooled) on the outside by liquid circulating through tubing in a helical configuration around the exterior of the cone. Two targets were built, one for the frequency range 18–110 GHz, and a smaller one for 60–230 GHz. Both targets operate between 80 and 350 K. The uncertainty in the standard's brightness temperature varies with frequency, ranging from 0.084 to 0.111 K.

The standard targets designed at NIST were intended for earth-based laboratory use. If a target is to be used in an application such as a satellite-based radiometer, the size and mass of the entire structure is a crucial consideration.

9.5 Target Reflectivity Effects

9.5.1 Effect of Target Reflectivity

In remote-sensing radiometry, calibration targets and the unknown target are generally viewed through the same antenna and measurement path between radiometer and antenna. Consequently, the complications introduced in Chapter 3 by different measurement paths (the different efficiencies and mismatch factors) are expected to be absent, and the radiometer equation for a total-power radiometer, Eq. (3.14), should reduce to

$$T_x = T_c + \frac{(p_x - p_c)}{(p_h - p_c)}(T_h - T_c) \tag{9.21}$$

or some equivalent form, as is commonly used in remote-sensing radiometry. There is, however, a potential complication.

As the name implies, remote-sensing targets of interest tend to be remote, located at a significant distance from the radiometer and antenna. On the other hand, calibration targets are often quite close to the antenna. Depending on the reflectivity of the calibration target, this can result in a different reflection coefficient and mismatch factor at plane 1 in Figure 9.2 when viewing the calibration target as opposed to the distant target of interest. Consequently, the effective input noise temperature of the radiometer may be different for the two cases, introducing an additional error and uncertainty. The antenna emits noise which then gets reflected back into the radiometer in the case of a nearby reflective

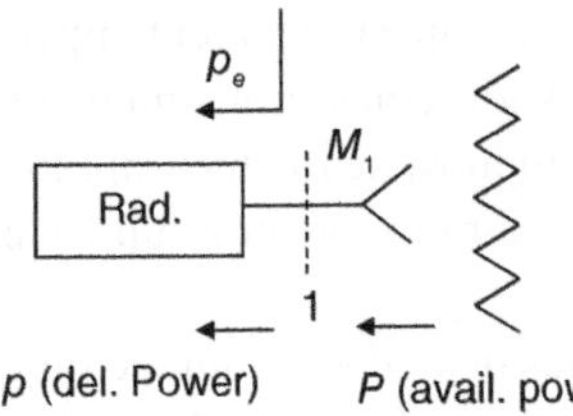

Figure 9.2 Antenna viewing calibration target. p_e represents the effective input noise power density of the radiometer, $p_e = k_B T_e$. Source: From Randa and Walker [10]/IEEE.

target. As was discussed in Chapter 3, if the radiometer has a front-end isolator, it is impervious to such changes, but if it is not isolated there will be some small effect. The effect should be small because calibration targets are designed to have high emissivity and low reflectivity, but it can be significant. In [10], the effect is estimated to be several tenths of a K or as large as a few K for two practical radiometer/antenna/target combinations. If the radiometers were isolated, the effect would be less than 0.1 K.

9.5.2 Measurement of Target Reflectivity

The reflectivity is a crucial property of a calibration target because of its direct link to the emissivity, through Kirchhoff's radiation law. It is usually determined by analysis of the design using measured material properties [8, 11–14]. It can also be measured, but the measurements are very challenging because calibration targets are designed to have the lowest reflectivity (highest emissivity) possible. A measurement method has been developed and applied, using measurements of the reflection coefficient of the antenna as a function of the distance between the antenna and the target [15].

The antenna reflection coefficient is measured by a vector network analyzer (VNA) attached to plane x in Figure 9.1, and the distance to the target is varied. The reflection coefficient exhibits a ripple pattern due to interference between the reflection from the antenna itself, which is independent of the separation distance, and the reflection from the target, which varies with the distance. The response was calibrated using measurements at different distances with a reflective metal plate in place of the target and with no reflective target at all, with the antenna radiating into an empty anechoic chamber. The method was verified using a polystyrene sheet with known reflectivity. Measurements with three different antennas were performed in order to check the results and to verify that they did not depend on details of the radiating antenna. For the calibration target tested, the reflectivity was consistent with zero, with an upper bound of about −40 dB.

References

1 F. Ulaby, R. Moore, and A. Fung, *Microwave Remote Sensing: Fundamentals and Radiometry*, Vols. **1–3**, Artech House, Norwood, MA, 1981.

2 J.P. Rice and B.C. Johnson, "NIST activities in support of space-based radiometric remote sensing," *Proceedings of SPIE* **4450**, pp. 108-126 (2001).

3 J. Randa, "Traceability for microwave remote-sensing radiometry," NIST Interagency Report, NISTIR 6631 (June 2004). https://tsapps.nist.gov/publication/get_pdf.cfm?pub_id=31906 (Accessed 17 March, 2022)

4 J. Randa, A.E. Cox, M. Francis, J. Guerrieri, and K. MacReynolds, "Standard radiometers and targets for microwave remote sensing," in *IGARSS*

2004 Conference Digest, Anchorage, AK, Paper 2TU_30_10 (September 2004). https://tsapps.nist.gov/publication/get_pdf.cfm?pub_id=31736 (Accessed 17 March, 2022).

5 J. Randa *et al.*, "Recommended terminology for microwave radiometry," NIST Technical Note 1551 (August 2008). https://doi.org/10.6028/NIST.TN.1551

6 D. Gu, D. Houtz, J. Randa, and D.K. Walker, "Extraction of illumination efficiency by solely radiometric measurements for improved brightness-temperature characterization of microwave blackbody target," *IEEE Transactions on Geoscience and Remote Sensing*, **50**, no. 11, pp. 4575-4583 (November 2012). https://doi.org/10.1109/TGRS.2012.2193890.

7 D. Gu, D. Houtz, J. Randa, and D.K. Walker, "Realization of a standard radiometer for microwave brightness-temperature measurements traceable to fundamental noise standards," in *2012 IEEE International Geoscience and Remote Sensing Symposium*, Munich, Germany (2012). https://doi.org/10.1109/IGARSS.2012.6350709.

8 D.A. Houtz, W. Emery, D. Gu, K Jacob, A. Murk, and D.K. Walker, "Electromagnetic design and performance of a conical microwave blackbody target for radiometer calibration," *IEEE Transactions on Geoscience and Remote Sensing*, **55**, no. 8, pp. 4586-4596 (August 2017).

9 D.A. Houtz, "NIST microwave blackbody: the design, testing, and verification of a conical brightness temperature source," University of Colorado Thesis (2017). https://scholar.colorado.edu/concern/graduate_thesis_or_dissertations/9w032318x (Accessed 17 March, 2022)

10 J. Randa, D.K. Walker, A.E. Cox, and R.L. Billinger, "Errors resulting from the reflectivity of calibration targets," *IEEE Transactions on Geoscience and Remote Sensing*, **43**, no. 1, pp. 50–58 (January 2005).

11 D.M. Jackson and A.J. Gasiewski, "Electromagnetic and thermal analyses of radiometer calibration targets," *IEEE Proceedings of International Geoscience And Remote Sensing Symposium*, Vol. **7**, Honolulu, HI, July 2000, 2827–2829.

12 A. Schröder, A. Murk, R. Wylde, D. Schobert, and M. Winser, "Brightness temperature computation of microwave calibration targets," *IEEE Transactions on Geoscience and Remote Sensing*, **55**, no. 12, pp. 7104-7112 (December 2017).

13 A. Schröder and A. Murk, "Numerical design and analysis of conical blackbody targets with advanced shape," *IEEE Transactions on Antennas and Propagation*, **64**, no. 5, pp. 1850–1858 (May 2016).

14 N. Feng and W. Wei, "The optimization design for microwave wide band blackbody calibration target," in *2008 International Conference on Microwave and Millimeter Wave Technology*, Nanjing, China, 2008, pp. 1695–1698. doi: 10.1109/ICMMT.2008.4540796.

15 D. Gu, D. Houtz, J. Randa, and D.K. Walker, "Reflectivity study of microwave blackbody target," *IEEE Transactions on Geoscience and Remote Sensing*, **49**, no. 9, pp. 3443–3451 (September 2011).

Index

a

adapters, measurements through 42, 43
ambient standard 12, 13
amplifier noise 47–82
APD *see* auxiliary passive device (APD)
asymmetry 36, 37
auxiliary passive device (APD) 112, 113
available gain 9, 47, 48
available power 7, 8
available-power ratio 8, 9
avalanche diode 19

b

BIPM *see* International Bureau of Weights and Measures (BIPM)
bipolar transistor 112
Bosma's Theorem 6, 7, 112
brightness 150
brightness temperature 147–159
  definition 150, 151
broadband mismatch error 38

c

calibration
  calibration target 152, 154–157
  noise source 40, 43, 44
  radiometer 24–28
  remote-sensing radiometer 149–154
CCEM *see* Consultative Committee for Electricity and Magnetism (CCEM)
check (of results) 29, 109–120
check standard 29
CIPM *see* International Committee for Weights and Measures (CIPM)
coaxial cryogenic standard 13–15
cold-attenuator method 123, 124
cold transfer standard 19
combined uncertainty 34, 39
common mode 139
Consultative Committee for Electricity and Magnetism (CCEM) 43
correlated errors 36, 37
cryogenic amplifiers 121–132
  matched noise-figure measurement 123–128
  noise-parameter measurement 128, 129
cryogenic standard 13–18

d

delivered power 7, 8
device under test (DUT) 23
Dicke radiometer 40, 41
differential amplifier 139–143
differential mode 139

digital radiometer 41, 42
DUT *see* device under test (DUT)

e

effective input noise temperature, definition 47, 50, 52, 53, 121–123
efficiency 8
ENR *see* excess noise ratio (ENR)
equivalent hot standard 18, 19
excess noise ratio (ENR) 5, 50

f

F_{min} 51
frequency offset error 38
frequency-variation method 66, 67, 129

g

Γ_{opt} 50
gain, available 9, 47, 48
gas-discharge tube 19
GTRF, Radiofrequency Working Group 43

h

high electron mobility transistor (HEMT) 100
hot (oven) standard 13

i

IF *see* intermediate frequency (IF)
illumination efficiency η_{IE} 153–155
inter-laboratory comparison 43, 44, 94–97
intermediate frequency (IF) 32
International Bureau of Weights and Measures (BIPM) 43
International Committee for Weights and Measures (CIPM) 43
intrinsic noise correlation matrix 6, 7, 135
isolator 26, 31

l

laboratory environment 23
limit, high or low frequency 2–4
liquid nitrogen (LN) 14–17
LNA *see* low-noise amplifier (LNA)
local oscillator (LO) 31
low-noise amplifier (LNA) 31, 75, 76, 125, 128
low-pass filter (LPF) 31
LPF *see* low-pass filter (LPF)

m

mismatch factor 8, 26, 27
Monte Carlo uncertainty evaluation 34, 71–77, 101–104
multiport amplifiers 133–146

n

National Institute of Standards and Technology (NIST) 13, 23, 29
National Measurement Institute (NMI) 43, 147
NFRad *see* noise-figure radiometer (NFRad)
NIST *see* National Institute of Standards and Technology (NIST)
NMI *see* National Measurement Institute (NMI)
NMOS *see* N-type metal oxide semiconductor (NMOS)
noise correlation matrix 6, 135, 136
noise factor, definition 47
noise figure 47
  measurement 49, 50
  multiport 136–138
noise-figure radiometer (NFRad) 29, 30, 32

noise matrix *see* noise correlation matrix
noise parameters, definition 50–55
  IEEE (circuit) representation 51
  measurement 55–67
  on-wafer measurement 88–100
  wave representation 52, 53
noise-parameter uncertainties 67–77
  on-wafer uncertainties 101–104
noise temperature, definition 4, 5, 48
    see also effective input noise temperature, definition
  measurement 23–46
  on-wafer measurement 85–87
  standards 11–22
  translation 9
N-type metal oxide semiconductor (NMOS) 91
Nyquist's Theorem 1, 2

o

on-wafer measurements 83–108
  noise parameters 88–107
  noise temperature 85–87
oven standard 13

p

physical bounds 54, 55, 111, 112
platinum resistance thermometer (PRT) 29, 155
primary standard 11, 12
PRT *see* platinum resistance thermometer (PRT)
pseudo waves 84

q

quantum factor 1, 2

r

R_n 51
radio frequency (RF) 31
radiometer 23
radiometer equation 25
  for isolated total-power radiometer 27, 28
Rayleigh-Jeans approximation 4, 11, 150
reference temperature, T_0 5
remote sensing 147–159
reverse noise temperature, T_{rev} 54, 91, 110, 111
RF *see* radio frequency (RF)

s

secondary standard 11, 19
signal-to-noise ratio 47, 138, 139
SI, international system 43
solid-state noise source 19
source-pull measurements 66
standard radiometer 149–154
standard target 148, 155
  design 155, 156
switching radiometer 40, 41
synthetic primary standard 19, 20

t

T_e *see* effective input noise temperature
$T_{e,min}$ 51
target reflectivity 156–157
tertiary standard 11
total-power radiometer 24–40
  design 29–32
  testing 32–34
traceability 43, 44, 147, 155
transfer standard 11, 19
transistor, measurements, see on–wafer measurements
  model-assisted measurements 95, 97–100
  reference planes 84, 85, 88–90, 94
T_0, reference temperature 5
tunable primary standard 18
two-tier calibration 89, 93, 94

type-A uncertainty 34–36
type-B uncertainty 34, 36–39

u
ultraviolet catastrophe 2
uncertainty
  noise parameters 67–77
  on-wafer noise parameters 101–104
  total-power radiometer 34–40

v
vacuum-fluctuation contribution 121–123
variable-termination unit (VTU) 75
vector network analyzer (VNA) 38, 50

w
water plate 23
waveguide cryogenic standard 15–18

x
X parameters, definition 52, 53

y
Y_{opt} 51
Y-factor method 49, 50

z
Z_0 51

Printed and bound by CPI Group (UK) Ltd, Croydon, CR0 4YY

16/04/2025

14658586-0004